CONTENTS

Preface 6

Chapter One 8

Chapter Two 26

Chapter Three 46

Chapter Four 70

Chapter Five 97

Chapter Six 125

Chapter Seven 157

Notes and Sources 189

Appendix A 195

Appendix B 202

Appendix C 210

About the Author 215

About The Kinnara Fund 217

The Biology of You 219

A Small Request 220

Acknowledgement 222

THE BIOLOGY OF YOU · BOOK ONE

Your DNA, Your Blueprint

Understanding Your Genome, Your Health Risks, and the Power You Hold to Shape Your Own Biology

Debra-Ann Smith

A Kinnara Fund Publication

Your DNA, Your Blueprint

First published 2026

Published by The Kinnara Fund

A 501(C) Nonprofit · Washington, D.C

TheKinnaraFund.Org

ISBN: 979-[8]-[9962419]-[0]-[3] (Paperback)

Cover design courtesy of The Kinnara Fund

Internal illustrations courtesy of The Kinnara Fund

Genetic risk information discussed in this book reflects the current state of scientific understanding as of the time of writing. The field of genomics advances rapidly; readers are encouraged to consult current clinical guidelines and up-to-date research sources for the most recent guidance.

Printed in United States of America

10 9 8 7 6 5 4 3 2 1

Father in Heaven, Thank you.

Preface

How to Use This Book

You do not need to be a scientist to read this book. You do not need any prior knowledge of biology, genetics, or medicine. You need only to be curious about your own health and willing to engage with ideas that are, genuinely, some of the most remarkable in modern science.

The Biology of You is an eight-book series. Each book takes one area of biotechnology and explores it from a single perspective: yours. What does this science mean for your health? What is currently available that you can access and use? What is coming that you should know about? And where does the science meet ethics — the questions about privacy, consent, equity, and responsibility that no conversation about biotechnology can responsibly avoid?

This first book, Your DNA, Your Blueprint, begins at the beginning: with the genome itself. It is, in many ways, the foundational volume. The biological principles introduced here — how DNA is inherited, how genes are expressed, what epigenetics means, how risk is calculated from genetic variants — underpin every other book in the series. As the series grows, each subsequent book can be read independently, but you will find the concepts easier and richer if you start here.

Every chapter in this book is written to be read straight through, but also to function as a reference — something you return to when your test results arrive, when a result concerns you, or when you need to explain something to a doctor or a family member.

A note on what this book is not: it is not medical advice. Nothing in these pages should substitute for a conversation with a doctor, a genetic counselor, or another qualified healthcare professional who knows your specific situation. What this book is, is preparation for those conversations — the background knowledge that transforms a clinical appointment from a bewildering encounter with unfamiliar terminology into a productive dialogue between people who understand each other.

Throughout each chapter you will find two types of callout box. Blue boxes contain explanations and key concepts — the building blocks you need to understand the section that follows. Teal boxes are action-oriented: they contain specific guidance, resources, and things you can do with the information you have just read. The boxes are designed to be useful on a second reading as well as a first — you can navigate by scanning them once you know the book.

Each chapter ends with a summary of the key ideas and a bridge to the next chapter. Chapter Seven closes with a complete action plan — fifteen specific steps across three phases — that brings together everything from the previous six chapters into a practical guide for using your genomic information over a lifetime.

The instruction manual has been inside you all along. This book teaches you to read it.

Debra-Ann Smith

Summer 2026

Chapter One

The Instruction Manual Inside You

Have you ever wondered why you got your mother's nose but your father's temper? Why your sister can eat anything and stay slim while you just have to look at bread? Why certain diseases seem to follow families like uninvited guests, generation after generation — and why other people in the same family escape them entirely?

The answer to all of these questions lives inside you right now. Not in a hospital, not in a textbook, not waiting to be discovered by someone else. Inside you. In every cell of your body, there is a complete biological instruction set — your own personal manual — that has been shaping you since the moment you were conceived.

This chapter is the beginning of learning how to read it.

We are not going to wade into difficult science. We are going to go on a short journey — one that starts with something as simple as the question of who you are, and ends with one of the most empowering ideas in modern medicine: that while you were born with a specific set of biological instructions, you have far more influence over how those instructions play out than most people ever realize.

You were not handed a fixed destiny.
You were handed a starting point
— and the story of your health
is one you are actively writing,
whether you know it or not.

Your Body Has a Language — and It Has Been Talking to Itself Your Entire Life

Imagine that on the day you were born, someone handed you a set of architectural plans for the most complex building ever designed. Not just the structure — but the plumbing, the electrical wiring, the heating system, the security, the maintenance schedule, and even the contingency plans for every possible kind of damage or threat. Every detail about what you would look like, how your organs would function, how your immune system would defend you, how you would age.

Now imagine that those plans were written in a language you had never been taught. They have been sitting there your whole life, quietly running everything, and nobody ever showed you how to read them.

That is DNA.

DNA — deoxyribonucleic acid — is the molecule that carries your biological instructions. It is written in a chemical alphabet of just four letters: A, T, G, and C. Those four letters, arranged in an enormously long sequence, make up your genome — your complete personal instruction set. Every living thing on Earth uses the same four-letter alphabet. What differs is the sequence. Your sequence is uniquely yours.

First, let's hold space for this thought: that sequence has been inside you from the very beginning. It was there when you were a single cell. It guided the construction of every organ, every nerve, every strand of hair. Right now, as you read this, it is still running — being read, copied, acted upon — in a continuous biological conversation that your body has been having with itself since before you took your first breath.

You have never been separate from your DNA. It is not a test you take at a lab. It is not something that belongs to scientists. It is the most intimate thing about you and understanding even a small piece of it, changes how you see your own health forever.

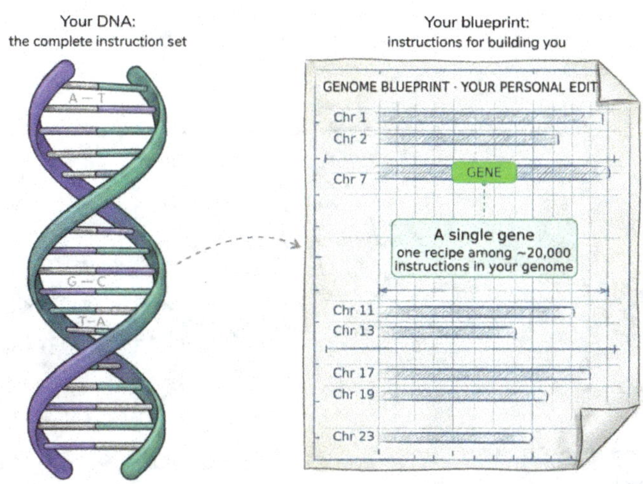

Figure 1: *DNA as architectural blueprint — a double helix showing the two backbone strands and their base-pair rungs, unfolding into a genome blueprint with one gene highlighted on Chromosome 7. The blueprint represents your complete genome — approximately twenty thousand genes, each carrying instructions for a specific biological task.*

Genes: The Individual Instructions Within the Manual

Within your genome, there are specific sections of that four-letter sequence that carry instructions for building particular things your body needs. These sections are called genes.

Think of a gene the way you might think of a recipe in a cookbook. The cookbook is your genome. A single recipe — say, the instructions for making a specific protein your immune system uses to fight infection — is a gene. You have somewhere in the region of twenty thousand of these recipes, each one doing something specific, each one part of the larger story of how your body functions.

Every one of the ~20,000 genes in your genome follows this same path — from written instruction to working molecule

Figure 2: *The cookbook metaphor — your genome is the complete cookbook; each gene is one recipe within it; each protein produced is the finished dish doing a real job in your body. The cookbook is your genome — the full library of everything your body knows how to make. A gene is one specific recipe. And the protein that gene produces is the finished dish: a working molecule doing a real job right now.*

Proteins are the workhorses of your body. They build your muscles, carry oxygen in your blood, send signals between your brain and your organs, defend you from bacteria and viruses, and regulate almost every chemical process keeping you alive right now. When a gene is working properly, it produces the right protein at the right time. When a gene carries a variation — a small difference in the sequence — that protein may behave differently. Sometimes that difference is harmless. Sometimes it matters a great deal.

This is why genetics can feel both deeply personal and slightly alarming when you first encounter it. Your genes are not abstract science — they are the instructions behind the things you feel, the conditions you are prone to, and the way your body responds to the world around you. A single small variation in a single gene is the reason some people process caffeine quickly and sleep soundly after an evening coffee, while others are wired until 3am. Another variation influences how your body responds to certain medications. Another shapes your likelihood of developing high blood pressure.

Figure 3: *Same coffee, different gene — one person asleep at 11pm, one still wide awake. The person on the left carries a CYP1A2 variant that metabolizes caffeine quickly: the coffee cleared their system hours ago. The person on the right carries a slower-metabolizing variant: the caffeine is still active, blocking the adenosine receptors that would normally signal sleep. Same drink. Same hour. Two completely different experiences — genetics in everyday life.*

A gene is not a verdict. It is a recipe — and like any recipe, what you do with the ingredients matters just as much as the instructions on the page.

This is not fate. This is information. And information, as we are going to discover throughout this book, is extraordinarily powerful when you know what to do with it.

This Is Where The Layers Begin To Peel Back:

You Are Not a Prisoner to Your Blueprint

Beyond the common headlines lies a more profound truth—a part of the story that changes the way you sit with your own history and the genes that shaped it.

You might assume that once you know what genes you have, the picture is complete. You have the manual, you know the instructions, and the outcome is more or less set. If heart disease runs in your family, you brace yourself for heart disease. If your mother had Type 2 diabetes, you assume yours is coming too. If the test says you carry a risk variant, you carry the risk — end of story.

But that is not how it works. Not even close.

Here is what scientists have discovered, and what is still reshaping medicine as you read this: having a gene — or even a high-risk version of a gene — does not automatically mean that gene is running at full volume in your body. Genes can be switched on or switched off. They can be turned up or turned down. They can be silenced for years and then activated. And a remarkable portion of what controls that switching is not fixed in your DNA at all. This is where a word you may have heard comes in: epigenetics.

Epi is a Greek prefix meaning 'above' or 'upon.' Epigenetics is, literally, the layer of control that sits above your DNA. It does not change the sequence of your genes — the letters stay the same. But it determines which genes get read loudly, which get read quietly, and which are effectively shelved for the moment.

Think of it this way:

Your genome is the full script for a play — every line that could ever be spoken is written there. Epigenetics is the director. The director does not rewrite the script, but decides which scenes are performed tonight, which actors are given the microphone, and which parts of the story stay in the background.

Two productions of the same play, directed differently, can feel like entirely different experiences. Two people with very similar DNA, living very different lives, can have very different health outcomes — for exactly this reason.

Now we reach the heart of the matter that should genuinely excite you: you have influence over the director.

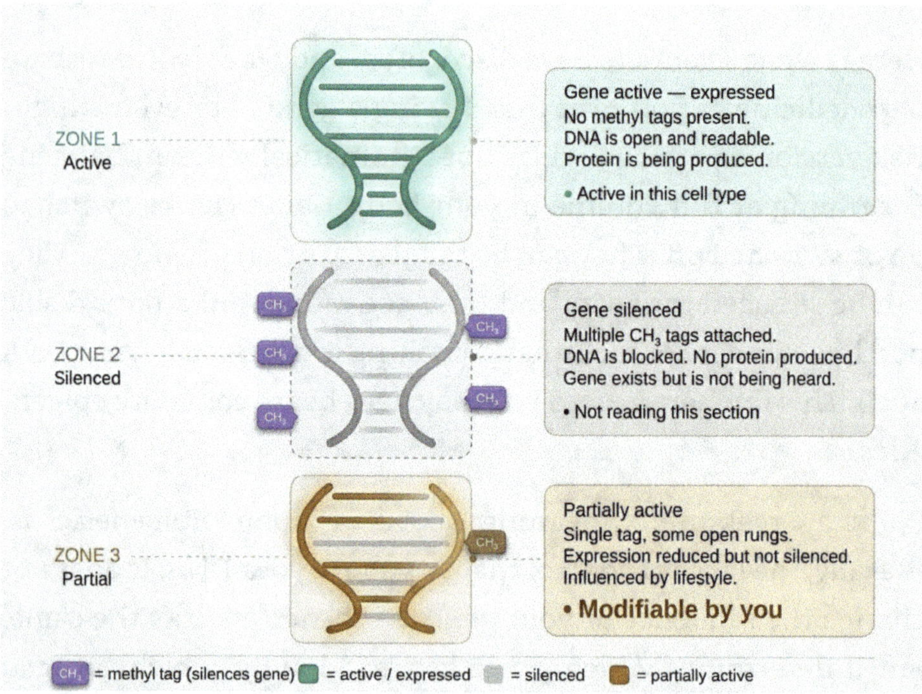

Figure 4: *The epigenetic layer — methyl tags (CH_3) silencing gene sections while others remain active. Zone 3 shows the modifiable middle ground where lifestyle plays a direct role.*

The diagram above shows three zones of the same DNA strand. In Zone 1 (teal), no chemical tags are present — the gene is open, readable, and actively producing a protein. In Zone 2 (gray), clusters of methyl tags have attached to the strand, physically blocking the reading machinery from accessing that section. The gene exists. It is simply not being heard. Zone 3 (amber) is the most important zone for your purposes: it is partially active, with only one tag present. This is the zone most directly influenced by the choices you make every day.

The food you eat leaves chemical signals that attach to your DNA and adjust which genes are being expressed. Regular physical activity changes the way genes involved in inflammation and metabolism are read in your muscles, your heart, and even your brain. Chronic stress — the kind that stays with you for months or years — alters the epigenetic landscape of genes involved in your immune system and your mood. Quality sleep allows your cells to reset and repair epigenetic patterns that have been disturbed during the day.

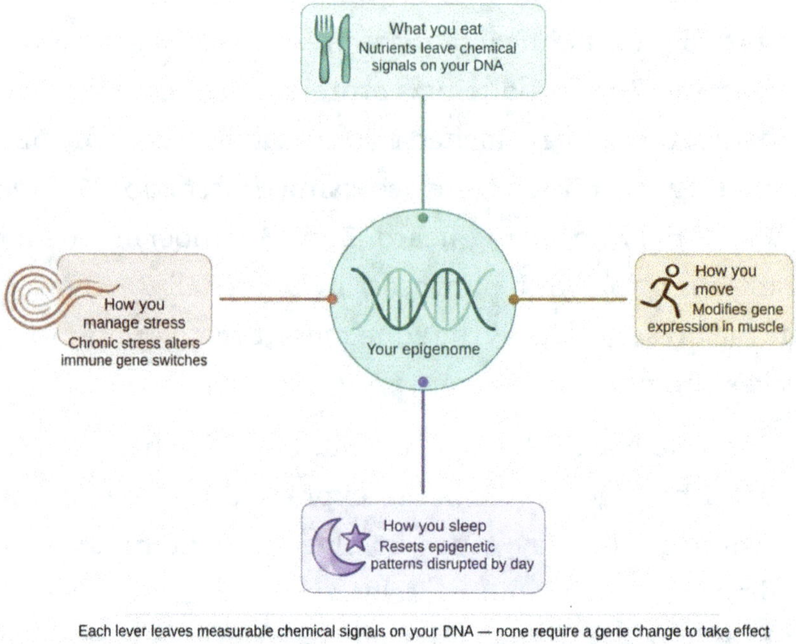

Figure 5: *The four lifestyle levers — what you eat, how you move, how you sleep, how you manage stress — each directly influences the epigenetic layer that sits above your DNA, affecting which genes are expressed and how loudly.* Commissioned for The Biology of You series.

None of this means that lifestyle alone can overcome every genetic challenge. Some conditions are genuinely driven by genetics in ways that require medical intervention, and we will look at those honestly throughout this book. But the picture is far more dynamic, far more hopeful, and far more personally relevant than the old model of 'genes equal destiny' ever allowed for.

In one landmark long-term study, researchers looked at people who carried high genetic risk variants for heart disease. Among those who also had what the researchers called an unfavorable lifestyle — poor diet, low activity, smoking — the risk played out almost exactly as the genetics predicted. But among the people with the same high-risk genes who maintained a healthy life-style, the actual incidence of heart disease was reduced by more than half. The genes had not changed. The expression of those genes — shaped by the way those people lived — had.

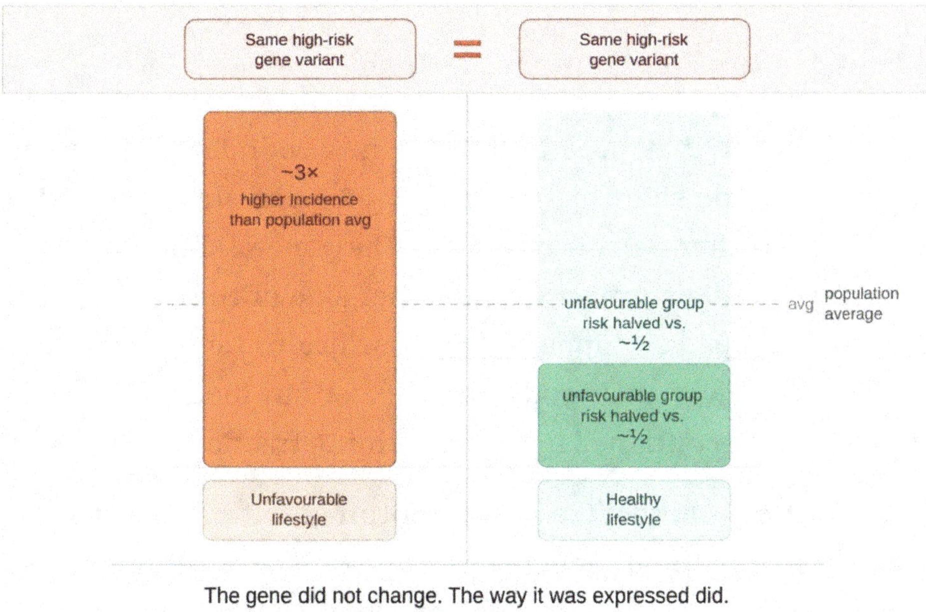

Figure 6: *Same high-risk gene variant, two different outcomes — the unfavorable lifestyle group shows approximately 3× higher disease incidence; the healthy lifestyle group cuts realized risk by roughly half. The genes did not change. The expression of those genes — shaped by how those people lived — did.*

Your DNA is real. Your risk variants, if you have them, are real. But they are not the whole story, and you are not a passive reader of a plot that has already been written.

> *Genetics loads the gun. The way you live — what you eat, how you move, how you sleep, how you manage stress — has a very real say in whether it is ever fired.*

What Makes You Genetically You

Here is a fact that puts all of this in a wonderfully human perspective: you share approximately 99.9 percent of your DNA with every other person on Earth. The genetic differences between you and a stranger on the other side of the world — the ones that shape your appearance, your health tendencies, your ancestry, the way your body handles certain foods or medications — are all contained in that remaining fraction of a percent.

That fraction sounds like almost nothing. But given the sheer length of the human genome, it still translates to millions of individual points where one person's sequence differs from another's. And those differences, accumulated across thousands of generations of human migration, adaptation, and survival, contain a remarkable amount of information about who you are biologically.

The most common type of variation is called a SNP — a single nucleotide polymorphism, though you will almost never need to use that phrase. A SNP (pronounced 'snip') is simply a position in your genome where different people carry different letters. You might have an A where someone else has a G. In the vast majority of cases, this makes no difference whatsoever — it is simply part of the beautiful variation that makes human beings diverse. But some SNPs sit in or near genes that have significant functions, and those are the ones that can influence things like disease risk, medication response, and physical traits.

When you take a consumer genetic test — like 23andMe or AncestryDNA — this is largely what it is scanning for: hundreds of thousands of these SNP positions across your genome, compared against large databases of people whose health outcomes are known. From that comparison, it calculates probabilities. Not certainties — probabilities. The distinction matters enormously, and we will spend a good portion of Chapter Two making sure you know exactly how to interpret what these tests are telling you.

For now, simply appreciate this: the tiny fraction of your DNA that makes you different from everyone else contains clues about your personal health that, until very recently in human history, nobody could access at all. The fact that you can now hold that information in your hands — and, more importantly, act on it — is one of the genuinely extraordinary developments of our time.

To put that in human terms: the next time you sit across from a family member who seems to share your health tendencies, you are looking at a version of that 0.1 percent made visible. The genes you carry in common are not coincidences. They are the biological thread connecting your histories — and, as we are about to see, your family is the oldest genetic test available to you.

* * *

Your Family History: The Genetic Test You Already Have

Before we go any further, there is something important to acknowledge. You do not need to spend a penny on a genetic test to begin understanding your biological inheritance. You already have access to the oldest and, in many ways, most reliable genetic tool available: your family history.

Think about what you know — or could find out — about the health of your parents, your grandparents, your siblings, your aunts and uncles. The conditions that have appeared repeatedly across generations in your family are not coincidences. They are patterns. They are your biology leaving a trail of breadcrumbs across time, telling you something about the genetic tendencies that have been passed down to you.

Heart disease appearing in multiple family members before the age of sixty. A grandmother and two aunts with breast cancer. A father and both his brothers with Type 2 diabetes. These are not just family stories — they are data. And unlike a polygenic risk score from a consumer test, your family history comes with context, with memory, with human meaning attached.

If you have never formally mapped your family medical history, that is the very first practical step we will encourage you to take before you do anything else in this book. It costs nothing, it can be assembled in an afternoon of conversations with relatives, and it can tell your doctor things about your risk profile that no off-the-shelf test can replicate.

We will return to this in Chapter Two, with a free downloadable template you can use. For now, plant the seed: your family is a living record of your genetics, and the stories they carry about their health are among the most relevant things you can know about your own.

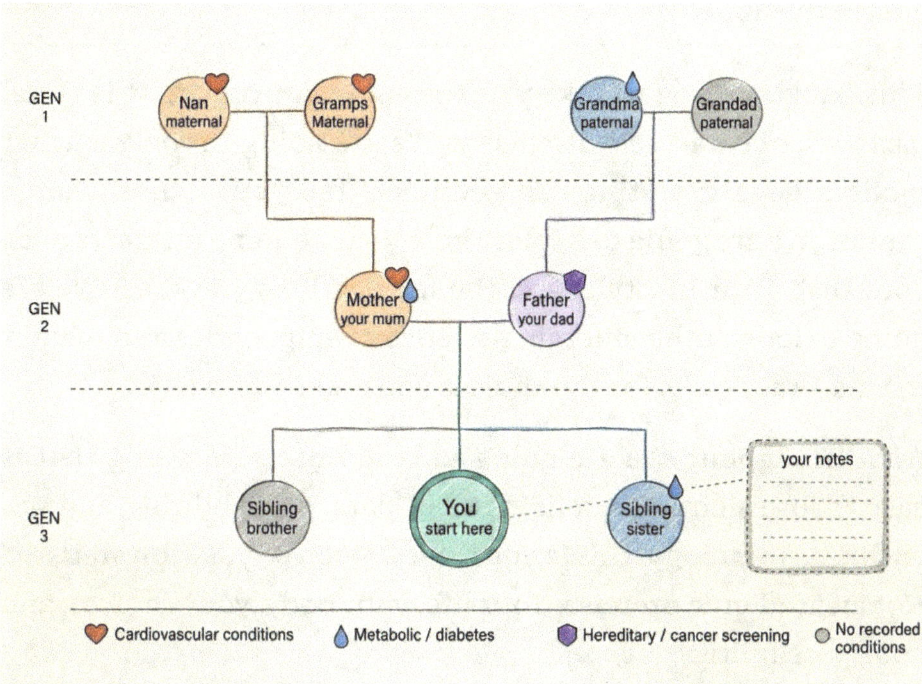

Figure 7: *A three-generation family medical history map — the oldest and most accessible genetic tool you already possess. Heart icons: cardiovascular. Drop icons: metabolic/diabetes. Shield icons: hereditary/cancer.*

The family tree above is a template, not a diagnosis. Notice how the maternal side carries a cluster of cardiovascular conditions (red hearts) across two generations — Nan, Gramps, and Mother all share that biological thread. The paternal grandmother's metabolic condition has appeared again in the younger sibling. These visible patterns across generations are exactly what a clinician looks for when assessing inherited risk. Your own version of this map, even a rough one jotted on a notepad, is one of the most valuable documents you can bring to a doctor's appointment.

The Adventure Ahead

This book is going to take you through a journey that is equal parts science and self-discovery. We are going to look at what modern genetic testing can genuinely tell you — and what it cannot. We are going to explore how your genome interacts with your diet, your lifestyle, and the medications you take. We are going to look at the emerging science of epigenetics and what it means, practically, for the choices you make every day.

And throughout, we are going to hold onto something that is easy to lose sight of in a field full of data, probabilities, and scientific terminology: this is your story. Not a population statistic. Not an academic exercise. Your life, your body, your choices, and your possibilities.

The instruction manual has been inside you all along. You are just learning, perhaps for the first time, how to read it.

> *The most powerful shift in modern medicine is not a new drug or a new technology. It is the moment a person looks at their own biology and thinks: now I understand what I can do.*

What We Covered in This Chapter

Before you turn the page, here are the ideas worth carrying forward:

- DNA is your complete biological instruction set — a unique sequence of four chemical letters that has been shaping your body since before you were born and is still actively at work right now.
- Genes are the individual instructions within that sequence — recipes for building the proteins that run every process in your body, from your immune response to your metabolism to your mood.
- Having a gene variant associated with a condition does not mean you will develop that condition. Genes are tendencies, not certainties.
- Epigenetics is the layer of biological control that sits above your DNA — influenced by what you eat, how you move, your stress levels, your sleep, and your environment. It determines which genes are active and how loudly they run.
- You share 99.9% of your DNA with all other humans. The small fraction that differs encodes your individual health tendencies, ancestry, and biological traits.
- Your family medical history is your oldest and most accessible genetic tool. Begin collecting it before you invest in any testing.
- This is not about fear. It is about knowledge — and the freedom that comes from understanding your own biology well enough to work with it.

Coming up in Chapter Two — Spitting in a Tube: How Consumer Genetic Tests Actually Work. We go inside the science of at-home DNA kits, explain what a risk score really means, compare the major testing companies, and give you a practical guide to reading your results with confidence and context.

+ + +

Chapter Two

Spitting in a Tube

How Consumer Genetic Tests Actually Work

Somewhere in a laboratory right now, a technician is processing a tube of saliva. The saliva came from a person who registered a kit online, spat into a small funnel, snapped a cap on a plastic vial, dropped it in a prepaid envelope, and forgot about it for three weeks. That person is expecting to receive something remarkable back: a window into their own biology that previous generations never had access to at all.

What actually happens between that tube leaving the post and the report arriving in their inbox is a journey that very few people know, and fewer still understand. And yet, understanding it — even just the shape of it, without the technical detail — is the difference between reading your results with confidence and reading them with the kind of low-grade anxiety that makes a useful tool feel like a threat.

This chapter is about demystifying that journey. We are going to follow your saliva from the moment it leaves your body to the moment it becomes data on your screen. And more importantly, we are going to spend time on the part that most test companies gloss over: what those results actually mean for you as a living, breathing person — and what they genuinely do not.

> *A genetic test does not read your future. It reads a set of tendencies — tendencies that are real and worth knowing, but that only tell part of the story of your health.*

What Is Actually in Your Saliva

Before we look at what the lab does with your sample, it helps to understand what you are actually sending them.

Saliva is mostly water. But floating in that water are cells — specifically, epithelial cells shed from the lining of your cheeks and mouth as a normal part of the body's constant process of renewal. Inside every one of those cells is a nucleus. And inside every nucleus is a complete copy of your genome — that full three-billion-letter instruction set we explored in Chapter One.

When you spit into a collection tube, you are collecting those cells. The tube contains a preserving solution that stabilizes the DNA inside them, keeping it intact for the days it will spend in transit to the laboratory. Some kits collect a cheek swab instead of saliva — the principle is identical, just a different way of gathering the same cells.

At the laboratory, the cells are broken open — lysed, in the technical language — and the DNA is extracted and purified. What remains is a solution containing your actual genome: the physical molecule that has been inside you since before you were born, now isolated, cleaned, and ready to be read.

It is one of those moments worth pausing on. Something as unremarkable as a few minutes of saliva contains, within the cells it carries, the complete biological instructions for you. The technology that allows us to read those instructions for a few hundred dollars is less than thirty years old. For the vast majority of human history, this information was entirely inaccessible — not just to individuals, but to science itself.

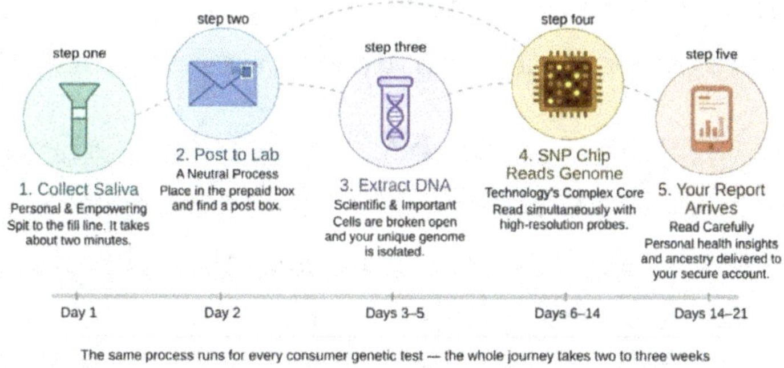

Figure 1: *From saliva to report — five steps: collect saliva, post to lab, extract DNA, SNP chip reads genome, report arrives. The whole journey takes two to three weeks.*
 Commissioned for The Biology of You series.

The Reading Machine: How Your DNA Gets Decoded

The next step — the actual reading of your DNA — is called genotyping, and it is where the real technology comes in.

You may remember from Chapter One that your genome is written in a four-letter alphabet: A, T, G, and C. Your complete genome contains around three billion of these letters. Reading all three billion in a single test is possible — it is called whole genome sequencing, and we will discuss it shortly — but most consumer tests do not do this. Instead, they use a more targeted approach called an SNP array.

An SNP is a specific position in the genome where people commonly differ from one another. One person might have an A at a particular position; another person might have a G. Most SNPs are entirely harmless, simply natural variation. But a subset of SNPs are located near or within genes with important functions, and these are the ones that consumer tests focus on.

A modern SNP array can check hundreds of thousands of these positions simultaneously — often between 650,000 and 900,000 specific spots across your genome. The chip itself is about the size of a credit card, covered in millions of tiny probes. Each probe is designed to bind to one specific DNA sequence. When your prepared DNA sample is washed over the chip, the relevant fragments stick to their matching probes. A laser scanner then reads the chip, detecting which probes have bound and which have not, and software translates this pattern into your genotype — the set of letters you carry at each of those hundreds of thousands of positions.

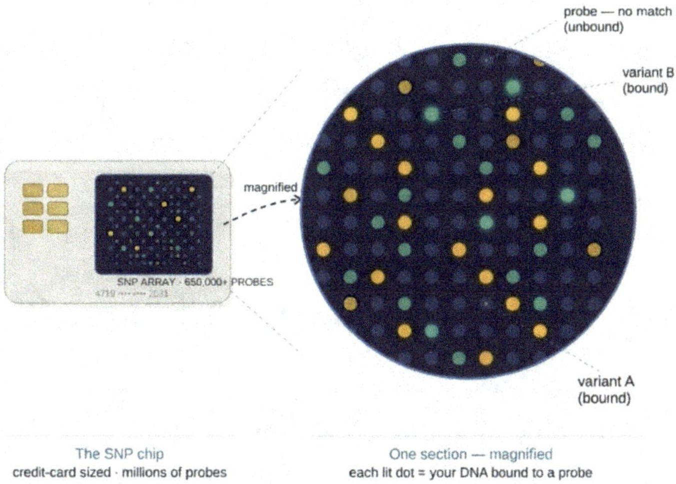

Figure 2: *DNA SNP chip concept — a credit-card sized microarray with a magnified circular inset showing a grid of probe dots, several glowing where DNA fragments have bound.*

The whole process, from saliva to raw genotype data, typically takes one to three weeks depending on the company and the volume of samples being processed.

Genotyping vs. whole genome sequencing

A SNP array checks specific known positions in your genome — the ones already linked to meaningful traits and conditions. Think of it as auditing the chapters of your instruction manual that scientists have already read and annotated.

Whole genome sequencing reads every single letter. It costs more and produces vastly more data — most of which, at present, scientists still do not know how to interpret. For most people, an SNP array provides more immediately actionable results for a fraction of the price.

The Human Genome Project completed the first full human genome sequence in 2003, at a cost of around three billion dollars. Today, both SNP array genotyping and whole genome sequencing cost under two hundred dollars. The bottleneck is no longer the sequencing — it is the interpretation.

From Raw Data to Your Report: The Art of Interpretation

The genotyping chip produces raw data — a list of hundreds of thousands of letter-pairs, one for each SNP position tested. On its own, this data tells you nothing useful. It is what the company does with that data next that produces the report you eventually read.

This is where understanding diverges sharply between companies — and where a great deal of the nuance around genetic testing lives.

The companies compare your SNP data against large scientific databases — primarily studies called genome-wide association studies, or GWAS. A GWAS takes a very large group of people, some with a particular condition and some without, and looks for SNP variants that appear significantly more often in one group than the other. If a specific variant is found, say, 30% more often in people who developed Type 2 diabetes than in those who did not, that variant gets flagged as associated with Type 2 diabetes risk.

The test company then checks whether you carry that variant — and if you do, your report notes an elevated risk for that condition. This sounds straightforward. But this is exactly the part that requires careful reading.

An association is not a cause. A variant appearing more frequently in people with a condition does not mean the variant itself causes the condition. It means the two things tend to occur together in the populations studied — and those populations have historically been heavily skewed toward people of European ancestry, which affects how applicable the findings are to people of other backgrounds.

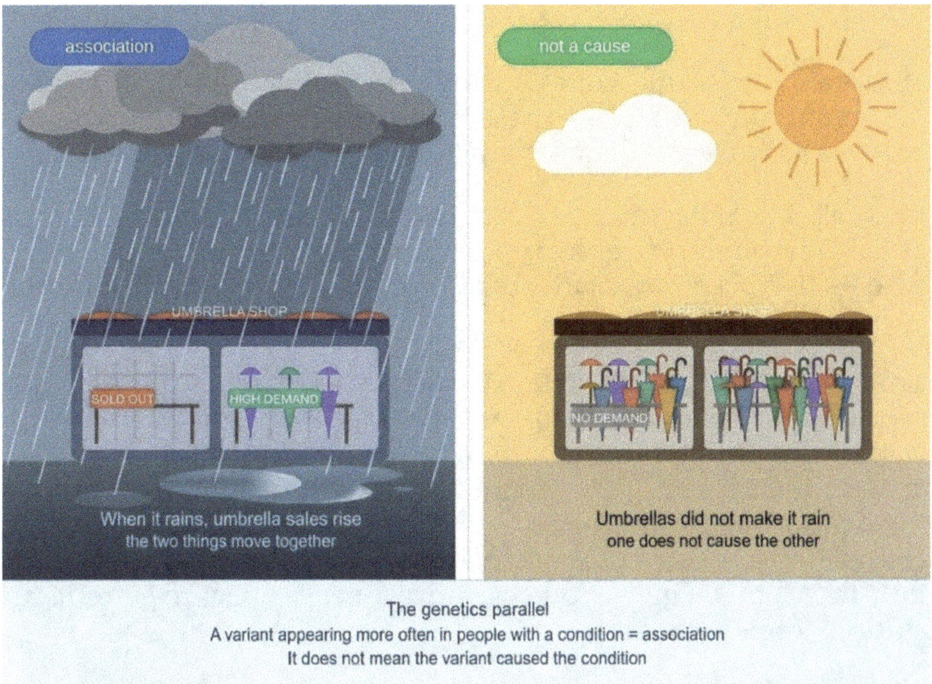

Figure 3: *Association is not cause — left panel: rain and high umbrella sales (labelled association); right panel: umbrellas in a dry shop (labelled not a cause). The genetics parallel: a variant appearing more often in people with a condition does not mean it caused the condition.*

What this means if you are not of European ancestry

As of 2025, people of European ancestry still account for more than 90 percent of all participants in the genome-wide association studies underpinning most consumer genetic tests. This is a real and acknowledged limitation of the science as it currently stands — and it affects every risk figure in your report.

It does not mean your results are worthless. It means they should be interpreted with more caution, and in context. Risk associations identified in predominantly European study populations may be stronger, weaker, or not yet detected in yours. Some variants common in your ancestry may not appear in the database at all.
Which results are most and least affected:

More reliable: single-gene findings such as BRCA1/2 and pharmacogenomic variants — these tend to transfer better across ancestries because the genes themselves are well-characterized.

Less reliable: polygenic risk scores for common conditions such as cardiovascular disease, type 2 diabetes, and Alzheimer's disease — these are most affected by database bias and should be treated as approximate guides rather than firm numbers.

This is changing. The NIH All of Us Research Program (allofus.nih.gov) has enrolled over 750,000 diverse participants across the United States and is now returning personalized genetic health results — including hereditary disease risk and medication-gene interactions — at no cost, with free genetic counseling available in English and Spanish. If you are in the US, joining All of Us gives you the most diversity-aware genetic results currently available.

Resources for non-European ancestry interpretation:

Find a Genetic Counselor — nsgc.org/findageneticcounselor (US) or ask your GP for an NHS genetics referral (UK). Request a counselor with experience in your specific ancestry background.

African ancestry: the H3Africa consortium produced landmark genomic data from 30 African countries — ask your counselor whether this data is relevant to your specific region of origin.

South and East Asian ancestry: the GenomeAsia 100K project has published significant variant data for South Asian and East Asian populations. Mention your specific heritage when discussing results.

All backgrounds: Color Health (color.com) offers equity-focused genetic counseling and clinical testing, including as the official counseling partner for the All of Us program.

See Chapter Three for specific guidance on how ancestral context affects the interpretation of BRCA, APOE e4, and cardiovascular risk scores.

Furthermore, most common conditions are not driven by a single variant. They are polygenic — influenced by hundreds or thousands of variants, each contributing a tiny sliver of risk or protection. The risk score in your report is typically the sum of many small effects, combined and weighted according to the available studies. It is a statistical estimate, not a biological verdict.

Reading a risk score on a genetic report is like reading a weather forecast. The probability is real and worth knowing. But the forecast is not the weather — and you still get to decide whether to carry an umbrella.

The Major Tests Compared: What You Are Actually Buying

The consumer genetic testing market is large and growing, but the landscape is more straightforward than it appears. Here is an honest picture of the main options and what distinguishes them.

23andMe is the most widely known consumer genomics company, and for good reason — they have one of the largest proprietary databases of genotyped individuals in the world, which allows them to produce increasingly refined ancestry and health reports. Their health reports are FDA-authorized in the United States for certain conditions including late-onset Alzheimer's disease risk (APOE variants), BRCA1/2 cancer risk variants, and several pharmacogenomic traits. Their ancestry reporting is particularly strong for people of European and Ashkenazi Jewish descent; less so for people from Africa, Asia, and the Caribbean, where reference populations are smaller.

AncestryDNA focuses primarily on ancestry rather than health. Their database is among the largest in the world — over twenty million people at the time of writing — which makes their ethnicity estimates and DNA matching for family discovery among the best available. They do not offer health reports directly, though your raw data can be downloaded and run through third-party interpretation tools.

MyHeritage DNA occupies similar ground to AncestryDNA, with strong family-tree integration and a large global database. Their health add-on service uses the same underlying genotype data to generate health and pharmacogenomic reports.

For people interested specifically in depth of data, companies like Nebula Genomics and Dante Labs offer whole genome sequencing at accessible price points — typically between one hundred and three hundred dollars. The tradeoff is that interpreting whole genome data requires considerably more scientific literacy, and the quality of their reporting interfaces varies widely.

A note worth making: most major consumer testing companies allow you to download your raw genotype file. This is a significant asset. Raw data from 23andMe or AncestryDNA can be uploaded to third-party interpretation services for additional analysis. We will cover this in Chapter Three.

Practical checklist: choosing a consumer test

- What is your primary goal — ancestry, health insights, medication guidance, or all three? Different companies do different things best.
- Check whether health reports are FDA-authorized (US) or CE-marked (Europe). This indicates regulatory review of the claims being made.
- Choose a company that allows you to download your raw genotype data. This future-proofs your investment — as science advances, you can re-interpret the same data with new tools.
- Consider the ancestry composition of the database. If your heritage is primarily African, South Asian, East Asian, or Caribbean, look specifically at which reference populations the company uses.
- Read the privacy policy before you order. Understand

whether your genetic data will be shared with third parties or used in research. Opting out of research sharing is usually possible — and worth doing if privacy matters to you.

What the Report Tells You — and What It Doesn't

When your report arrives, it will likely cover several categories of information. Understanding what each type of finding means — and where its limits are — is the key to using it wisely.

Notice that in every row, what the test measures and what it can tell you is genuinely useful. And in every row, there is something it cannot tell you — a gap between the data and the lived reality. That gap is not a flaw in the technology. It is an honest reflection of where genetics ends and where everything else — your choices, your environment, your epigenome — begins.

Reading a Risk Score Without Losing Your Mind

The part of a genetic report that causes the most anxiety is, almost universally, the health risk section. You open the report expecting curiosity and instead find a number that says your risk of a particular condition is elevated — and it sits with you in a way that is hard to shake.

Here is a framework for reading risk scores that will serve you every time you encounter one.

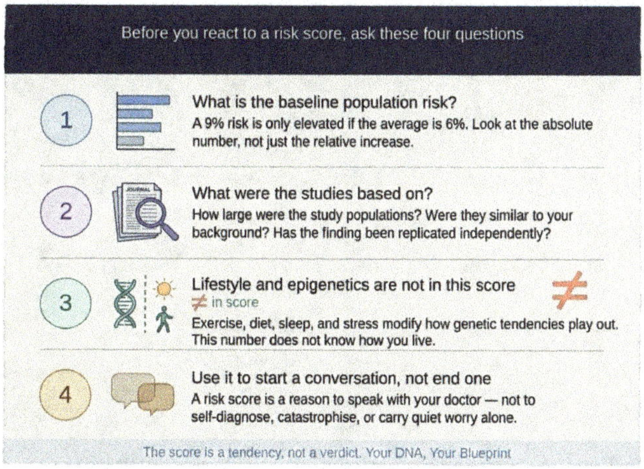

Figure 4: *Four questions to ask when reading a risk score: 1. What is the baseline population risk? 2. What were the studies based on? 3. Lifestyle and epigenetics are not in this score. 4. Use it to start a conversation, not end one.*

First: always ask what the baseline risk is. A risk score only makes sense in relation to the average. If the general population has a 6% lifetime risk of developing a particular condition, and your score puts you at 9%, your risk is fifty percent higher than average. That sounds alarming. But it means 91 out of every 100 people with your genetic profile will not develop that condition. The absolute risk is what matters for your decisions — not the relative increase.

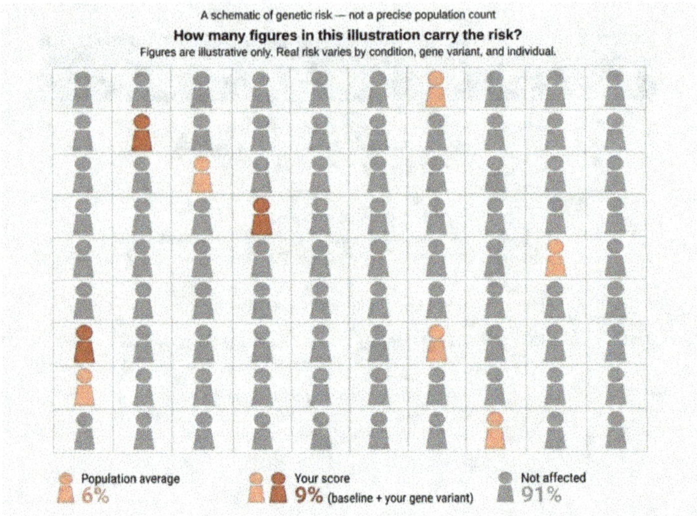

Figure 5: *A schematic of genetic risk — not a precise population count. 6 coral figures show baseline population risk; 9 total highlighted show an elevated personal score; 91 grey figures are not affected in this illustration. Figures are illustrative only.*

Second: ask what the risk score is actually based on. How large were the studies? Were the populations studied similar to your background? Has the finding been replicated across multiple independent studies, or does it rest on a single GWAS? The better companies give you this context in the report. If they do not, it is a legitimate question to raise with a healthcare provider.

Third: remember that a risk score is a snapshot of genetic tendency, not a measure of your current health. It does not account for the lifestyle and epigenetic factors we discussed in Chapter One. It does not know whether you exercise, what you eat, whether you smoke, how you sleep, or how you manage stress. All of those things are real and powerful modifiers of how genetic tendencies actually play out in a life.

Fourth: use the score as a starting point for a conversation, not as a conclusion. A finding worth paying attention to is a reason to speak with your doctor — to discuss what monitoring, screening, or lifestyle adjustments might be appropriate given your profile. It is not a reason to self-diagnose, catastrophize, or carry a quiet worry you never act on.

A word on findings that feel heavy

Some findings carry more weight than others. The APOE e4 variant, associated with elevated Alzheimer's risk, and the BRCA1/2 variants, associated with elevated breast and ovarian cancer risk, are in a different category from most consumer test findings — not because they are certainties, but because they have strong clinical implications that warrant professional guidance.

If you receive a result in either of these categories, please do not process it alone. A genetic counselor — a healthcare professional specifically trained in interpreting genetic information and its personal and family implications — can help you understand what the finding means, what screening options exist, and how to discuss it with relatives who may also be affected.

23andMe provides extensive support resources alongside these findings. You can also ask your GP for a referral to a clinical genetics service. The finding is not the end of a conversation — it is the beginning of one.

The Privacy Question You Should Not Skip

Before we close this chapter, there is a conversation that most consumer testing guides quietly avoid. Your genetic data is the most personal information you will ever generate. It is unique to you, permanent, and biologically shared — to varying degrees — with every relative you have. Once it exists in a database, you cannot un-create it.

This is not a reason not to test. It is a reason to test thoughtfully.

The major consumer genomics companies all have privacy policies, but they vary considerably. Some default to sharing anonymized genetic data with research partners unless you opt out. Some have entered data-sharing agreements with pharmaceutical companies. Some cooperate with law enforcement requests under certain conditions. Some store your raw genotype data indefinitely; others allow you to delete it on request.

There are also practical financial considerations. In some countries — including the United States — there are legal protections against genetic discrimination in health insurance and employment under the Genetic Information Nondiscrimination Act, known as GINA. But these protections have gaps: life insurance, disability insurance, and long-term care insurance are not covered. If you are in the market for these products, understanding the landscape in your jurisdiction before you test is worth a short piece of research.

For most people, the benefits of genetic self-knowledge outweigh these risks — particularly if you take the simple step of opting out of data sharing and downloading your raw data to keep locally. But making that decision with full awareness is considerably better than making it by default.

Your DNA belongs to you. The decision about who else gets access to it — and on what terms — is yours to make. Make it deliberately.

Your First Practical Step

If you have not yet taken a consumer genetic test and are considering it, here is a straightforward process for getting started on the right foot.

Before you order anything, spend thirty minutes writing down what you already know about your family's medical history — parents, grandparents, siblings, aunts and uncles. A full six-section template for recording three generations of family medical history is available free to download at thekinnarafund.org/blueprint, and is reproduced in full as Appendix A at the back of this book. Even a partial record — conditions you know about, ages of diagnosis, causes of death — is far more useful than no record at all.

Free download — Your Family Medical History template

The full six-section printable template (this book's Appendix A) is available as a free PDF download.

↗ **thekinnarafund.org/blueprint**

NOTE: The template is also reproduced in full as Appendix A at the back of this book.

Decide on your primary goal. If ancestry is your main interest, AncestryDNA or MyHeritage offer the most extensive databases.

If health insights are your priority, 23andMe currently offers the most comprehensive FDA-authorized consumer health reports.

Check for current promotions. These companies run frequent sales — sometimes discounting tests by thirty to fifty percent. There is rarely a reason to pay full retail price.

When ordering, read the consent screens carefully. Make an active choice about research data sharing, data storage, and family matching features. These defaults are set in the company's interest, not necessarily yours.

When your results arrive, read the introductory material before you dive into the health sections. Fifteen minutes of context saves a great deal of unnecessary worry.

Download your raw data file immediately and store it somewhere safe. This is a digital asset — its interpretive value will increase as the science advances, and you may want to re-analyze it with tools that do not yet exist.

If you have already taken a test and your results are sitting in an app you opened once and never returned to, this book is an invitation to go back. You have data about yourself that is genuinely worth understanding — and with the right context, it is considerably less alarming and considerably more useful than it may have felt the first time you read it.

Figure 6: *A DNA test kit on a kitchen table beside a morning coffee — the most powerful health tool you can order online fits in your letterbox. What matters is knowing how to read what comes back.*

What We Covered in This Chapter

Here are the ideas to carry forward:

- Consumer genetic tests work by collecting epithelial cells from your saliva, extracting your DNA, and scanning hundreds of thousands of specific positions across your genome using a technology called SNP array genotyping.
- Results are produced by comparing your genetic variants against large scientific databases called genome-wide association studies. These identify statistical associations between variants and health conditions — not causes.

- Risk scores are probabilities, not predictions. Always look at the absolute risk figure alongside any relative increase and ask what the baseline population risk is before deciding what a result means for you personally.
- The major consumer tests differ in their strengths: 23andMe for health insights, AncestryDNA and MyHeritage for family discovery, and whole genome sequencing services for depth of data.
- Your raw genotype file is a long-term asset. Download it, store it securely, and know that its interpretive value will grow as science advances.
- Some findings — particularly APOE e4 and BRCA1/2 variants — carry clinical weight that warrants professional guidance from a genetic counselor, not just self-managed reading.
- Your genetic data is permanent and has privacy implications. Read the terms of any service carefully, consider opting out of data sharing, and understand the legal landscape in your jurisdiction before you test.

Coming up in Chapter Three — Reading the Results with Confidence: a practical guide to interpreting your polygenic risk scores, understanding what your ancestry composition actually tells you, navigating pharmacogenomics, and knowing exactly when and how to take a finding to a doctor.

+ + +

Chapter Three

Reading the Risk

Disease Predisposition — Probability, Not Destiny

There is a particular kind of stillness that happens when you open a genetic report and see a result you were not expecting. Something inside you goes quiet. The screen is doing nothing unusual — it is just showing you text, the same as any other page — but the words are not like other words. They feel heavier. They feel like they are about something that cannot be undone.

If you have ever had that moment, or if you are anticipating it, this chapter is for you.

We are going to look closely at the three categories of genetic finding that carry the most weight in a consumer health report: cancer risk variants, particularly BRCA1 and BRCA2; Alzheimer's disease risk, centered on the APOE e4 variant; and cardiovascular disease predisposition, shaped by hundreds of genetic influences across the genome. These are the findings that matter most, that generate the most questions, and that are most frequently misunderstood.

The goal of this chapter is not to minimize what these results mean. Some of them are genuinely significant and warrant real action. The goal is to give you the knowledge and the framework to engage with them honestly — neither dismissing what is real nor catastrophizing what is not. Because the difference between knowing your risk and understanding your risk is the difference between paralysis and power.

Knowing your genetic risk is not the end of a conversation. It is the beginning of one — between you, your biology, and the choices still ahead of you.

Before We Begin: How Genes and Disease Actually Relate

Before we look at specific conditions, it helps to hold a clear mental model of the different ways genes relate to disease — because not all genetic findings carry the same kind of meaning.

At one end of the spectrum, there are rare single-gene conditions where a specific variant in a specific gene has a direct, high-probability link to a disease. Huntington's disease is the clearest example: if you inherit the relevant variant, the disease will almost certainly develop, typically in midlife. These are called monogenic conditions. They are the exception, not the rule, and most consumer tests do not screen for them without specific clinical indication.

At the other end of the spectrum — where most consumer health findings live — are common, complex diseases like Type 2 diabetes, heart disease, and most cancers. These conditions are influenced by hundreds or thousands of genetic variants, each contributing a tiny amount of risk or protection. No single variant is deterministic. They operate more like a weather system than a light switch: many small forces combining to push probability in one direction or another, against a backdrop of environment, lifestyle, and epigenetics that shifts throughout your life.

In the middle sits a small but important category: variants that carry genuinely elevated risk — not certainty, but a meaningful enough probability to change what you do. BRCA1 and BRCA2 sit here. Understanding this middle ground, clearly and honestly, is what this chapter is about.

Three categories of genetic risk

High penetrance, single gene — variant strongly predicts disease outcome. Examples: Huntington's disease, classical cystic fibrosis, familial hypercholesterolaemia. Rare. Usually not screened by consumer tests without clinical indication.

Moderate to high penetrance — variant meaningfully elevates risk but does not guarantee disease. Examples: BRCA1/BRCA2 variants for cancer, APOE e4 for Alzheimer's. Actionable. Warrants professional guidance.

Polygenic risk — hundreds of small-effect variants combined into a score. Examples: heart disease, Type 2 diabetes, depression. Common. Highly modifiable by lifestyle.

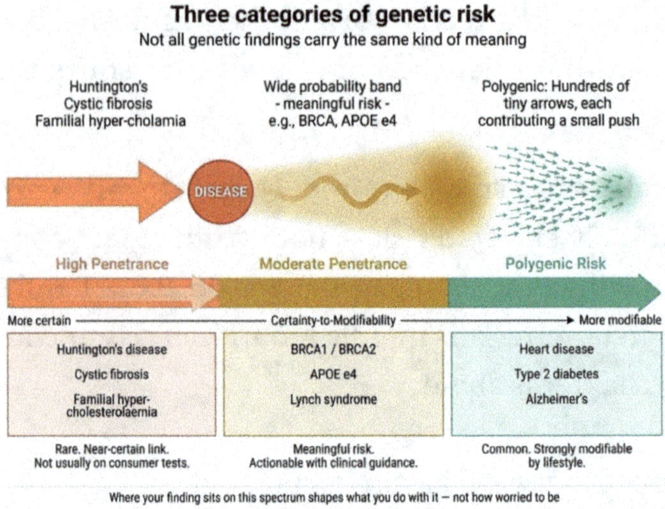

Figure 1: *Three categories of genetic risk — a horizontal spectrum from high-penetrance single-gene variants (left, coral) through moderate-penetrance variants like BRCA1/2 and APOE e4 (center, amber) to polygenic risk across hundreds of variants (right, teal). The spectrum runs from more certain but rarer (left) to more common but more modifiable (right).*

BRCA1 and BRCA2: What They Are, What They Do, and What a Positive Result Actually Means

BRCA is an abbreviation of BReast CAncer. BRCA1 and BRCA2 are genes that every person carries — two copies of each, one from each parent. In their normal state, these genes produce proteins that help repair damaged DNA. They are part of your body's built-in quality-control system, patrolling the genome and correcting errors before they become problems.

When a person inherits a harmful variant in one of these genes, that repair mechanism is compromised. The affected copy of the gene no longer produces a fully functional protein. If the second copy is later lost or damaged through normal cellular wear — which can happen over decades of cell division — the repair system fails. This creates conditions in which cancer-causing errors can accumulate unchecked.

This is why BRCA variants are associated with cancer: they do not cause it directly. They reduce your body's ability to catch and correct the kind of genetic damage that, over time, gives rise to it.

The numbers are significant and worth understanding clearly. In the general female population, the lifetime risk of developing breast cancer is around 12 to 13 percent. Among women who carry a harmful BRCA1 or BRCA2 variant, that lifetime risk rises to somewhere between 55 and 72 percent by age 70 to 80. For ovarian cancer, the general population lifetime risk is approximately 1 to 2 percent, while BRCA1 carriers face a risk of 39 to 46 percent and BRCA2 carriers face a risk of 11 to 29 percent.

These are meaningful numbers. They deserve to be taken seriously.

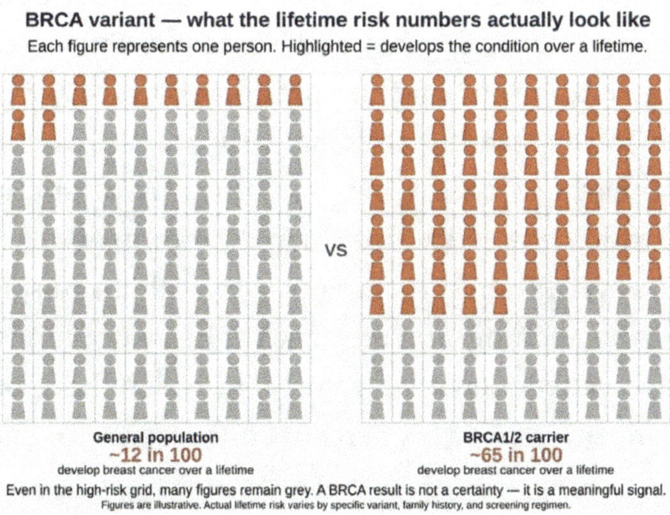

Figure 2: *BRCA variant — two grids of figures side by side: general population (~12 highlighted) vs BRCA carrier (~65 highlighted). Even in the high-risk grid, the majority of figures remain unhighlighted.*

Figures are illustrative only.

They also deserve context that is equally clear. A lifetime risk of 60 percent for breast cancer means that four in every ten women who carry the variant will not develop breast cancer. It means the outcome is not written. And crucially — in a way that distinguishes BRCA from a condition like Huntington's disease entirely — the knowledge of carrying a BRCA variant opens doors to action that can change those probabilities significantly.

A BRCA result is not a cancer diagnosis. It is specific, actionable information that gives you a meaningful head start on one of the most treatable diseases in medicine when caught early.

Women who know they carry a BRCA variant have options that women who do not know they carry it do not. Enhanced screening — alternating MRI and mammography every six months, beginning in the late twenties or early thirties — catches tumors at far earlier stages, when outcomes are dramatically better. Risk-reducing medications, including tamoxifen, can lower breast cancer risk in BRCA carriers. Surgical options, including preventive mastectomy and removal of the ovaries and fallopian tubes before cancer develops, reduce the risk of both breast and ovarian cancer by 85 to 95 percent in women who choose this path.

None of these choices is easy, and none of them should be made without careful conversation with a specialist. The point is not that any particular option is right — it is that options exist. And you cannot access them until you know.

A few additional facts worth carrying with you. BRCA variants are not a women's issue. Men who carry a harmful BRCA2 variant face significantly elevated lifetime risks of prostate cancer — up to 61 percent by age 80, compared with around 11 percent in the general population — as well as elevated risks of pancreatic cancer and, at a lower but still notable rate, male breast cancer. BRCA affects every carrier, regardless of sex.

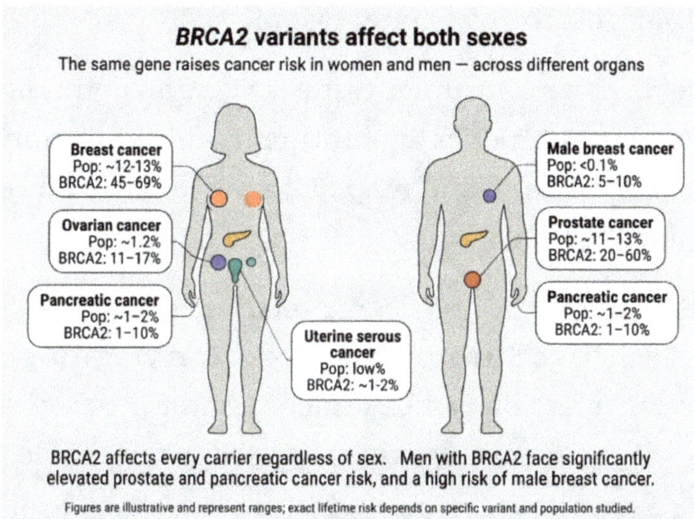

Figure 3: *BRCA variants affect both sexes — two stylized silhouettes showing cancer risk markers for breast, ovarian, prostate, and pancreatic cancers, with population versus BRCA-carrier risk percentages for each. The same gene raises risk in women and men across different organs.*

BRCA variants are also more common in certain ancestral populations. Among people of Ashkenazi Jewish descent, one in every forty individuals carries one of three specific BRCA1 or BRCA2 founder variants, compared with roughly one in three hundred to one in eight hundred in the general population. If your family heritage includes Eastern European Jewish ancestry, this is something your doctor should know.

If your consumer test shows a BRCA flag

Do not try to process this alone. Book an appointment with your GP and ask for a referral to a clinical genetics service or genetic counselor. Consumer tests screen for a limited number of known BRCA variants — a clinical test reads the entire gene and may find variants the consumer test missed.

A positive flag on a consumer test is a starting point, not a final answer. A clinical genetics appointment will confirm the result, assess the full picture, and map out a personalized monitoring or management plan.

Consider telling your first-degree relatives — parents, siblings, children. Each has a 50 percent chance of carrying the same variant. That conversation can be difficult, but it is one of the most genuinely life-saving things this knowledge can enable.

Many countries have dedicated hereditary breast and ovarian cancer clinics. In the UK, these run through NHS genetics services. In the US, the FORCE charity (Facing Our Risk of Cancer Empowered) provides extensive resources and community support for BRCA carriers.

Heart Disease: The Polygenic Story — and Why It Is the Most Hopeful of All

Heart disease is the leading cause of death worldwide. It is also, in the context of genetic risk, one of the most encouraging stories in modern medicine — because the genetics of cardiovascular disease are among the best-studied in existence, and the evidence that lifestyle and medication can dramatically modify genetic risk is overwhelming.

Unlike BRCA, there is no single gene variant that largely determines your cardiovascular fate. Heart disease risk is shaped by more than two thousand genetic variants, each contributing a small nudge toward or away from risk. Your consumer test will typically report this as a polygenic risk score — a number reflecting where you sit in a distribution of genetic risk relative to the general population.

The top eight percent of genetic risk scorers are estimated to have a likelihood of coronary artery disease more than three times higher than the rest of the population — a risk comparable to that seen in familial hypercholesterolaemia. People in this group tend to reach the threshold of ten percent cumulative heart disease risk more than fifteen years earlier than people in the lowest risk tier.

Those are significant findings. They are also findings that respond powerfully to intervention.

Research involving over a million individuals has shown that people in the high genetic risk tier who adopted a healthy lifestyle — no smoking, regular physical activity, maintaining a healthy weight, and a broadly heart-healthy diet — reduced their actual incidence of heart disease events by approximately half compared to high-risk individuals who did not. The genetic risk remained exactly the same in both groups. The biological expression of that risk diverged sharply.

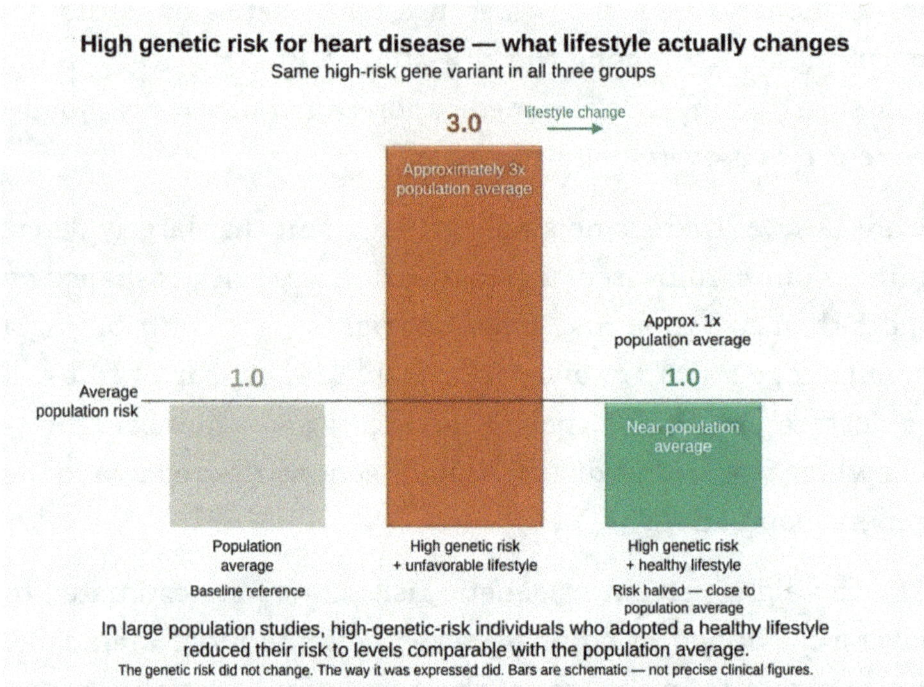

Figure 4: *Heart disease and genetic risk — three bars: population average, high genetic risk with unfavorable lifestyle, high genetic risk with healthy lifestyle. The third bar returns close to the population baseline, illustrating that high inherited risk is powerfully modified by how you live.*

Bar heights are schematic — not precise clinical figures.

Research involving over a million individuals has confirmed that people in the high genetic risk tier respond powerfully to active cardiovascular risk management. In clinical trials, participants with the highest inherited cardiovascular risk who received active treatment showed substantially larger absolute reductions in ten-year cardiovascular event risk than those with lower genetic risk receiving the same treatment. The genetic score was doing two things at once: identifying who was most at risk and identifying who would benefit most from acting on that risk. This is precision medicine in its most practical form: directing the most effective interventions toward the people who will benefit most from them. Discuss your cardiovascular polygenic score with your GP to determine what active risk management looks like in your specific situation.

A high polygenic score for heart disease is not a prophecy. In study after study, it is precisely the people with the highest genetic risk who gain the most from lifestyle change and early treatment.

The key genes involved in cardiovascular risk include those influencing LDL cholesterol levels — particularly the LDLR, APOB, and PCSK9 genes — as well as variants affecting blood pressure, inflammation, and the tendency of blood to clot. Some of these have individual significance: a harmful variant in LDLR underlies familial hypercholesterolemia, and carriers are often identified through family history. Your test may flag any of these individually, in addition to presenting an overall polygenic risk score.

What this means practically: if your report shows elevated cardiovascular genetic risk, your response should not be dread. It should be a conversation with your GP about your lipid profile, your blood pressure trend, and whether your current lifestyle places you in the group that halves their risk or the group that realizes it. The information is directional. It is telling you where to focus, not what to accept.

A note on cardiovascular risk and ancestry. In populations of African descent, the dominant driver of cardiovascular disease is hypertension rather than elevated LDL cholesterol. This is a well-established finding from large studies including the Heart of Soweto cohort, which showed that Black African patients had significantly lower LDL cholesterol and triglycerides than other groups, yet comparable or higher rates of cardiovascular disease — driven primarily by blood pressure. What this means in practice: if you are of African descent and your consumer genetic test returns a high cardiovascular polygenic score, the most important conversation to have with your GP is about your blood pressure trajectory, not your cholesterol level alone. The genetic risk is real; the primary pathway through which it expresses itself may differ from what the standard European-derived risk model assumes.

The same lifestyle levers that reduce cardiovascular risk broadly are particularly relevant here — and they work directly on blood pressure, not only on cholesterol. Regular aerobic exercise is among the most robustly supported interventions for blood pressure reduction: 150 minutes per week of moderate-intensity activity produces clinically meaningful falls in both systolic and diastolic pressure.

Chronic stress is an independent driver of sustained hypertension — the physiological stress response raises cortisol, which raises blood pressure, and when that response becomes the background state rather than a temporary one, the cardiovascular cost compounds over years.

Stress reduction practices — whether structured relaxation,

regular physical activity, improved sleep quality, or reduced workplace and financial pressure — are not soft recommendations. They address a concrete biological pathway. Good sleep quality and adequate sleep duration are independently associated with lower blood pressure and better cardiovascular outcomes across all ancestries, and the effect is particularly pronounced in individuals with high genetic cardiovascular risk. This does not change the advice to speak with your GP — it reinforces it. The conversation about your blood pressure trajectory is most useful when it includes an honest review of these lifestyle factors alongside any clinical management.

Alzheimer's Disease and the APOE Gene: The Most Misunderstood Result in Consumer Testing

Of all the findings a person can receive on a consumer genetic test, the APOE e4 result generates more anxiety — and more misinterpretation — than almost any other. It is worth spending time here, because understanding what APOE e4 actually means for most people is genuinely reassuring in ways that media coverage rarely manages to convey.

The APOE gene comes in three common variants, known as e2, e3, and e4. The e3 variant is the most common and is considered neutral with respect to Alzheimer's risk. The e2 variant is associated with a modestly lower-than-average risk. The e4 variant is the one that matters for risk discussions.

Approximately one in four people carries at least one copy of APOE e4. Most will never develop Alzheimer's disease. This is worth reading twice, because it runs counter to the way APOE e4 is usually discussed.

Having one copy of APOE e4 roughly doubles to triples the lifetime risk of Alzheimer's disease compared to people carrying two copies of e3. That sounds alarming. But the general population lifetime risk of Alzheimer's, for a person who reaches age 65, is approximately 15 percent. Doubling or tripling that risk gives a lifetime risk of roughly 25 to 30 percent for a single e4 copy carrier — meaning that more than seven in ten people with one copy of APOE e4 will not develop Alzheimer's disease. The majority outcome, even with this variant, is not the disease.

Two copies of APOE e4 — inherited from both parents — is a different situation. Only 2 to 3 percent of the population carry this genotype. Research published in 2024 in Nature Medicine suggested that people who are homozygous for APOE e4 may experience a distinct form of the disease biology, with over 95 percent showing Alzheimer's-related biomarkers by their mid-fifties. This is a genuinely significant finding and represents a category of risk that warrants specialist guidance.

For the single-copy carrier — which describes roughly one in four people — the picture is meaningful but not dire. APOE e4 is a risk factor, not a diagnosis. It cannot tell you whether you will develop Alzheimer's. It cannot tell you when. And it does not change many of the clinical recommendations your doctor would otherwise make, because there is currently no approved medication that can prevent late-onset Alzheimer's in someone who carries APOE e4 but has not yet developed symptoms. What it does do is point you toward lifestyle factors that the evidence suggests are genuinely protective.

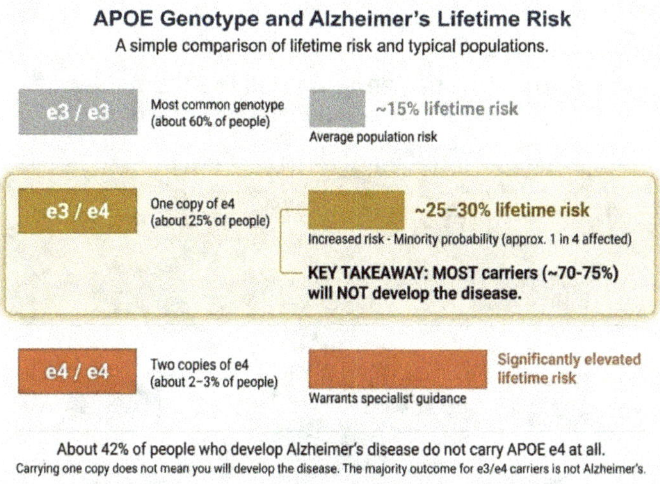

Figure 5: *APOE genotype and Alzheimer's lifetime risk — three rows showing e3/e3 (approximately 15% lifetime risk, most common), e3/e4 (approximately 25-30%, one copy), and e4/e4 (significantly elevated, two copies). The majority of e3/e4 carriers will not develop Alzheimer's disease.*

About 42 percent of people who develop Alzheimer's disease do not carry the APOE e4 variant at all. And most people who do carry it will never develop the disease. The variant matters — but it does not decide.

The modifiable risk factors for Alzheimer's disease — the things within your sphere of influence regardless of your genetic profile — are substantial. A landmark 2020 report in The Lancet identified twelve modifiable risk factors that together account for around 40 percent of all dementia cases globally. These include physical inactivity, social isolation, excessive alcohol use, smoking, obesity, high blood pressure, hearing loss, depression, diabetes, low educational engagement, traumatic head injury, and air pollution. None of these is a genetic factor. All of them are addressable.

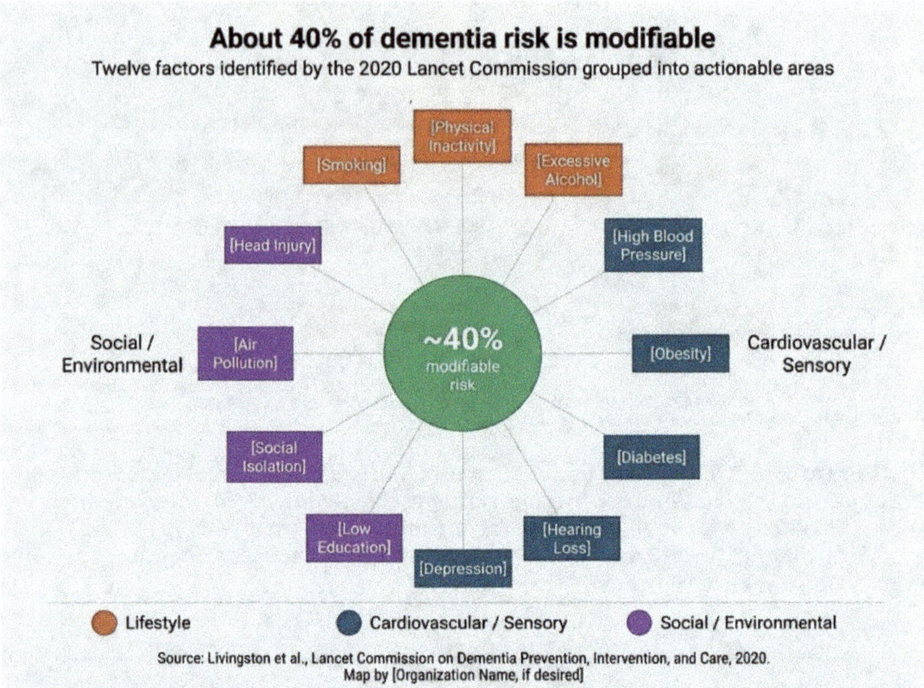

Figure 6: *About 40% of dementia risk is modifiable — twelve factors from the Lancet Commission 2020 arranged radially: lifestyle, cardiovascular/sensory, and social/environmental clusters. Source: Livingston et al., Lancet, 2020.*

For APOE e4 carriers in particular, cardiovascular risk management is especially relevant: APOE e4 also raises the risk of cardiovascular disease, and cardiovascular health and brain health are deeply intertwined. A carrier who manages their blood pressure, stays physically active, maintains social engagement, sleeps well, and avoids smoking is doing more to protect their brain than almost any pharmaceutical intervention currently available.

APOE e4 and the Chris Hemsworth effect

In 2022, the actor Chris Hemsworth disclosed publicly that genetic testing had revealed he carries two copies of APOE e4. Media coverage was immediate and somewhat catastrophizing — headlines suggesting he faced near-certain Alzheimer's disease.

This was not an accurate representation of the science. Even among those with two APOE e4 copies, the disease does not develop in everyone, and research into interventions for this group is active and advancing. More importantly, Hemsworth's disclosure prompted him to make significant lifestyle changes — improved sleep, reduced work schedule, enhanced fitness — that are precisely the kinds of modifications the evidence supports.

His story illustrates both the risk of genetic information without context, and its genuine value when used wisely: it changed his behavior in directions that are likely to be genuinely protective. That is exactly what genetic knowledge is for.

Other Disease Predispositions Worth Knowing

BRCA, cardiovascular risk, and APOE are the findings that receive the most attention, but a comprehensive consumer report will typically include information on a broader range of conditions. Here is an honest orientation to some others you may encounter.

Type 2 diabetes is a polygenic condition with well-established genetic contributors. Consumer tests can give you an elevated or reduced risk score, but the modifiable factors — weight, diet, physical activity, sleep quality — have such powerful influence over whether genetic risk translates into disease that a high score is best read as a prompt to take preventive lifestyle seriously, rather than as a warning of inevitable diagnosis.

Prostate cancer has both polygenic risk components and a specific moderate-to-high penetrance gene — BRCA2, which raises prostate cancer risk in men substantially. If a report flags elevated polygenic prostate cancer risk alongside a BRCA2 variant, that combination warrants a conversation with a doctor about when to begin PSA screening.

Coeliac disease has a strong genetic component: most people with coeliac disease carry one of two genetic variants in the HLA region of the genome, known as HLA-DQ2 and HLA-DQ8. Consumer tests often screen for these. Importantly, carrying these variants does not mean you have coeliac disease — the majority of people who carry them do not. But a positive result combined with any digestive symptoms makes an antibody test well worth pursuing.

Mental health conditions, including depression and schizophrenia, have real polygenic components that consumer tests increasingly report on. These results require particular care. The polygenic risk scores in this area are less predictive than cardiovascular or cancer risk scores, gene-environment interactions are complex and not yet well understood, and stigma around mental health makes these findings especially prone to misinterpretation. Treat them as background context rather than a clinical finding and speak with a professional before drawing conclusions.

Rare recessive conditions — carrier status for cystic fibrosis, sickle cell disease, spinal muscular atrophy, and others — are reported separately from disease risk, and their implications differ. Being a carrier for a recessive condition does not generally affect your own health. It becomes clinically relevant when planning to have children, because if both parents carry a variant in the same gene, each child has a one-in-four chance of inheriting two copies and developing the condition. Knowing your carrier status before pregnancy — while there is still time to consider options with a genetic counselor — is one of the most genuinely actionable things a consumer test can reveal.

The Question Nobody Asks Enough: What Do I Do Now?

Reading a risk result is one thing. Deciding what to do with it is another, and this is the part that most articles, most test reports, and most headline summaries leave frustratingly vague.

Here is a practical framework for moving from result to action, whatever the finding.

For a finding in the high-penetrance category — BRCA1/2 or another single-gene condition with well-established clinical implications — the first step is clinical confirmation. A consumer test screens for known variants; a clinical genetic test reads the full gene. Book an appointment with your GP, describe what the report showed, and ask for a referral to a genetics service. Do not make significant medical decisions based on a consumer test result alone.

For a finding in the moderate-risk category — a single APOE e4 copy, an elevated cardiovascular polygenic score — the primary response is a review of your modifiable risk factors with a clinician who understands the context. Ask for a lipid panel if you have not had one recently. Discuss your blood pressure. Have an honest conversation about your lifestyle. Use the genetic information as a compass, not a countdown.

For findings in the polygenic category — Type 2 diabetes risk, overall cancer risk scores — think of the information as directional motivation. Is this condition one that runs in your family? If so, that convergence of evidence is meaningful. If your family history is clean and the genetic risk is modest, you have less to be concerned about than the headline risk percentage might suggest.

For findings that feel confusing, disproportionately distressing, or that you are not sure what to do with — that is exactly what genetic counselors are trained for. They are healthcare professionals who specialize in interpreting genetic information in the context of a real human life, not a database entry. Asking to see one is not an overreaction. It is using the healthcare system precisely as it should be used.

A simple decision tree for your genetic results

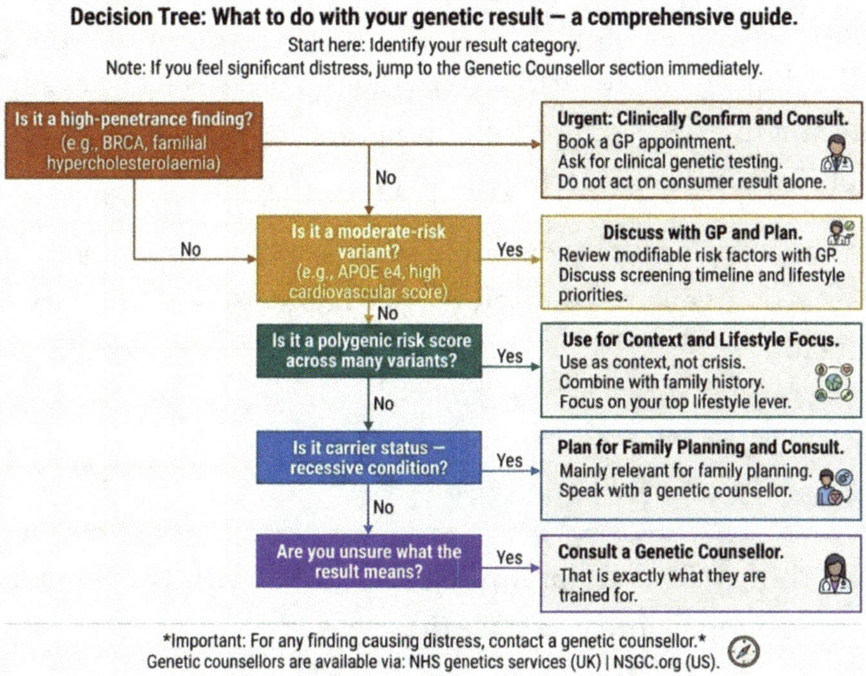

Figure 7: *What to do with your genetic result — a five-row decision guide linking each result category to a clear action pathway: high-penetrance finding, moderate-risk variant, polygenic risk score, carrier status, and uncertain or distressing finding.*

Does the finding relate to a high-penetrance condition (BRCA or another single-gene condition with established clinical implications)? → Book a GP appointment and ask for clinical genetic testing. Do not act on the consumer result alone.

Does the finding relate to a moderate-risk variant (single APOE e4, elevated cardiovascular polygenic score)? → Review your modifiable risk factors with your GP. Focus on the lifestyle changes with the most evidence. Consider screening timelines.

Does the finding relate to a polygenic risk across many variants? → Use it as context, not crisis. Combine it with your family history. Identify the one or two lifestyle levers with the highest impact for that specific condition.

Is the finding about carrier status for a recessive condition? → Relevant mainly for family planning. Speak with a genetic counselor if you are considering pregnancy.

Are you unsure what the finding means or how seriously to take it? → Ask a genetic counselor. That is precisely what they are there for.

Probability Is Not Destiny

Everything in this chapter points to a single idea that demands our reflection before you close these pages.

The biological instructions inside you are real. The variants that elevate your risk of certain conditions are real. The statistical differences between carrying BRCA2 and not carrying it, between having two copies of APOE e4 and none, between sitting in the top decile of cardiovascular polygenic risk and the bottom — all of it is real and all of it matters.

But probability is not destiny. A 60 percent lifetime risk of breast cancer means a 40 percent chance of never developing it — and means that knowing about it early gives you and your clinical team the clearest possible opportunity to act on that knowledge. A doubled risk of Alzheimer's disease in a population where the base risk is 15 percent still means that most carriers never develop the disease — and points toward a set of modifiable factors that the evidence supports with unusual consistency. A high polygenic cardiovascular score in study after study is associated with the greatest relative benefit from lifestyle change and active clinical management.

The most useful way to hold genetic risk information is not as a verdict but as a map. A map does not tell you what will happen. It tells you the terrain — where the difficult ground lies, and where the paths are. What you do with that knowledge is still, as it has always been, yours to decide.

> *The most empowering thing your genetics can give you is not a prediction. It is a reason to act — earlier, smarter, and with more information than any generation before you has ever had.*

What We Covered in This Chapter

Here are the ideas to carry forward:

- Genetic risk falls into three broad categories: high-penetrance single-gene variants (rare, strongly predictive), moderate-penetrance variants like BRCA and APOE e4 (significant but not certain), and polygenic risk across many small-effect variants (common, and highly responsive to lifestyle).
- BRCA1 and BRCA2 variants meaningfully raise lifetime risk of breast, ovarian, prostate, and pancreatic cancers. A positive consumer test result is a starting point for clinical genetic testing, not a final answer. Knowing is genuinely actionable — earlier screening and preventive options can dramatically change outcomes.
- Heart disease is the most polygenic of the common

conditions, and the evidence that lifestyle change and early lipid-lowering treatment reduce realized risk in high genetic risk individuals is among the strongest in genomic medicine. For people of African descent, blood pressure management is the primary cardiovascular intervention signal — and this applies regardless of what the polygenic score shows for cholesterol.

- APOE e4 increases Alzheimer's risk but is carried by roughly one in four people, most of whom will never develop the disease. The modifiable factors — cardiovascular health, physical activity, sleep, social connection — are within your control regardless of genotype.
- Other findings worth understanding include polygenic diabetes and prostate cancer risk, HLA variants for coeliac disease, and carrier status for recessive conditions — the last of which is particularly important for family planning.
- For high-penetrance findings, seek clinical genetic testing and specialist guidance. For moderate-risk findings, review modifiable factors with your GP. For polygenic findings, use the result as context and motivation. For anything confusing, ask a genetic counselor.
- Probability is not destiny. Every category of genetic risk discussed in this chapter is modified by factors you can influence. The point of knowing your risk is to do something useful with it.

Coming up in Chapter Four — Your Genes and Your Medicine Cabinet: Pharmacogenomics. Why the same medication can work brilliantly for one person, barely at all for another, and how knowing your genetic drug-metabolism profile is one of the most immediately practical applications of everything you have now learned.

✦ ✦ ✦

Chapter Four

Your Genes and Your Medicine Cabinet

Pharmacogenomics — Why One Drug Works for You but Not Your Neighbor

Imagine you and your neighbor are both prescribed the same blood pressure medication by the same doctor. Same drug, same dose, same instructions on the packet. A month later, you check in with each other. Your blood pressure is beautifully controlled. Your neighbor's hasn't shifted at all. Neither of you did anything wrong. The medication isn't defective. The doctor made a perfectly reasonable choice for both of you.

The reason the outcomes diverged is sitting quietly in your DNA.Your body processes medications through a complex network of enzymes — molecular machines that break drugs down, activate them, transform them, and eventually clear them from your system. The genes that build those enzymes come in different versions. And depending on which version you inherited, you might process a drug two times faster than your neighbor, or four times slower. You might activate a prodrug efficiently while they activate it barely at all. You might reach a toxic concentration on a dose that barely touches them.

This is pharmacogenomics: the study of how your genetic profile shapes your response to medications. And it is one of the most immediately practical — and most underused — pieces of information your DNA can offer you.

Unlike the risk scores for heart disease or Alzheimer's that we explored in the previous chapter, pharmacogenomic findings are not about probability over a lifetime. They are about what is happening in your body right now, every time you take a pill. They are about why certain antidepressants worked and others didn't. Why a pain medication left you groggy and strange while your friend took the same thing and felt perfectly fine. Why your warfarin dose seems to need constant adjustment while your neighbor has been stable for years. And unlike the disease risk associations in Chapter Three — which were derived predominantly from genome-wide studies of people of European ancestry — the mechanism at the heart of pharmacogenomics is biochemical and universal. CYP enzymes exist in every human being. The variants that make them faster or slower have been studied across all major population groups. This is, in many ways, the most ancestry-inclusive field in consumer genomics.

This chapter is about making those mysteries visible.

> *Your genes don't just influence whether you get a disease. They influence whether the treatment for that disease will work — and whether it will be safe.*

The Liver's Secret: How Your Body Processes Drugs

To understand pharmacogenomics, it helps to understand what

happens to a medication once it enters your body. Most people assume drugs simply travel to where they are needed and do their job. The reality is considerably more complicated — and considerably more dependent on your genetics.

When you swallow a tablet, it dissolves in your digestive system and its active compounds enter your bloodstream. From there, most drugs are carried to the liver — your body's primary chemical processing plant. In the liver, a family of enzymes called cytochrome P450 enzymes — the CYP enzymes — get to work. Their job is to metabolize the drug: to chemically transform it so that either it becomes active (some drugs are prodrugs, inactive until the liver processes them), or so that it can be safely eliminated from the body.

Different CYP enzymes handle different drugs. CYP2D6 handles roughly a quarter of all commonly prescribed medications. CYP2C19 handles dozens more, including several blood thinners and antidepressants. CYP2C9, CYP3A4, and others each have their own portfolios. And each of these enzymes is built from a gene that comes in multiple variants — versions that produce enzymes ranging from highly efficient to barely functional to entirely absent.

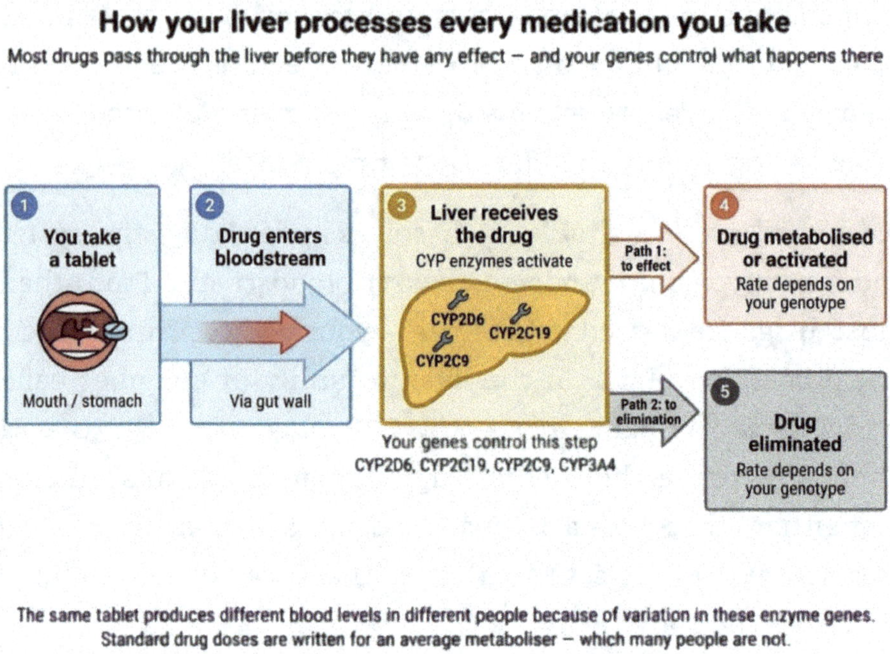

Figure 1: *The liver drug processing pipeline — five steps from tablet to effect or elimination, with the liver's CYP enzymes highlighted as the step controlled by your genes. Key enzymes: CYP2D6, CYP2C19, CYP2C9, CYP3A4. The same tablet produces different blood levels in different people because of variation in these enzyme genes.*

The implications unfold in four directions, depending on which variant you carry.

If you are a poor metabolizer of a particular enzyme, your body breaks down drugs in that enzyme's portfolio very slowly. Drugs that need to be eliminated accumulate — they build up to higher concentrations in your blood than intended, and you experience stronger effects and more side effects at doses that would be perfectly fine for someone else. If the drug is a prodrug that requires that enzyme to be activated, a poor metabolizer gets little or no benefit at all.

If you are an intermediate metabolizer, you process the drug more slowly than average but not negligibly — you may need a lower dose than standard to avoid side effects, or a higher dose of a prodrug to achieve the intended effect.

If you are a normal metabolizer — the category most drugs are designed for — the standard dose produces the expected outcome. This is the assumption every prescription is written against, even though a significant portion of the population does not fit it.

If you are an ultra-rapid metabolizer, you have extra copies of the gene, producing more enzyme than average. Your body clears drugs so quickly that a standard dose may not build to sufficient concentration to work. For prodrugs, the opposite is true: ultra-rapid metabolizers can activate a prodrug so quickly that dangerous concentrations build up almost immediately.

Four metabolizer types — and what they mean for your prescriptions

Poor metabolizer — enzyme works slowly or not at all. Drugs accumulate. Side effects increase. Prodrugs — medications that only become active after your body converts them — may not activate. Around 5–10% of people of European descent are CYP2D6 poor metabolizers.

Intermediate metabolizer — enzyme works at reduced capacity. Dose adjustments may be needed. Often the most common variant in some populations for certain genes.

Normal metabolizer — enzyme works as expected. Standard

doses are appropriate. Drug labeling and clinical trials are designed around this group.

Ultra-rapid metabolizer — extra gene copies, faster-than-normal processing. Drugs may be cleared before they take effect. Prodrugs may reach toxic levels quickly. CYP2D6 ultra-rapid metabolism affects approximately 1–2% of people of European descent but up to 29% of people of North African heritage.

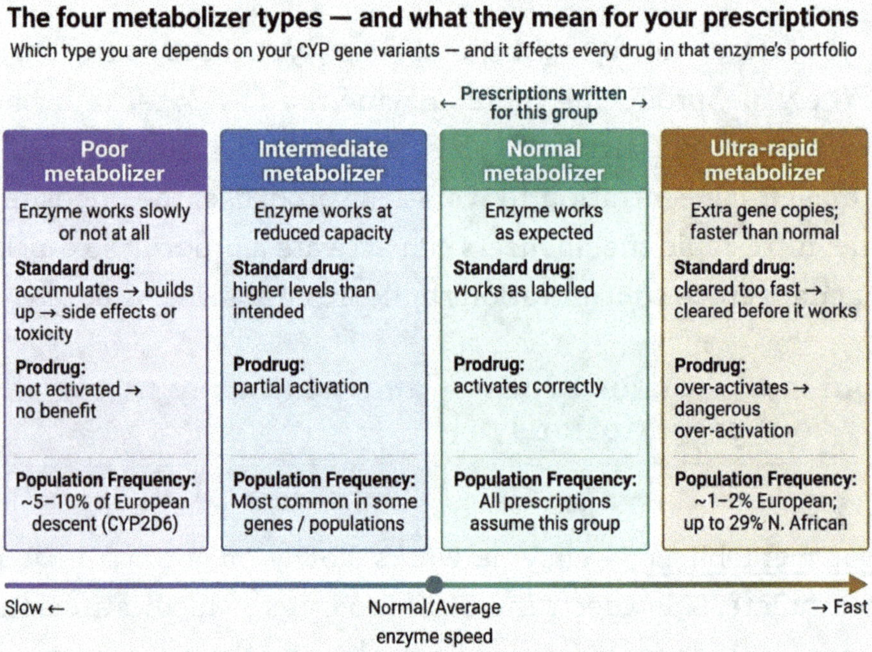

Figure 2: *The four metabolizer types — poor, intermediate, normal, and ultra-rapid — showing what happens to a standard drug and a prodrug at each position on the spectrum. All prescriptions are written for the normal metabolizer. The frequency of each type varies significantly by ancestry.*

The Codeine Story: When a Common Painkiller Becomes a Different Drug in Different People

The story of codeine is one of the clearest illustrations of why pharmacogenomics matters — not as an abstract scientific concept, but as a genuine matter of safety.

Codeine is a prodrug. When you take it for pain relief, it arrives in your body in an inactive form. It is only when the CYP2D6 enzyme converts approximately 10 percent of it into morphine that it produces its pain-relieving effect. The whole system is designed around the assumption that this conversion happens at a predictable rate.

But for a poor metabolizer of CYP2D6 — someone whose enzyme barely functions — that conversion either does not happen or happens at a fraction of the expected rate. These patients take codeine and feel nothing. Or nearly nothing. They are often labeled as drug-seekers or difficult patients when they report that the medication isn't working, when the true explanation is written in their DNA.

For an ultra-rapid metabolizer, the danger runs in the opposite direction. CYP2D6 ultra-rapid metabolizers convert codeine to morphine extremely quickly and in far greater quantities. In adults, this can cause dangerously high morphine levels — respiratory depression, extreme sedation, and in rare cases, death. In breastfeeding mothers who are ultra-rapid metabolizers, codeine can pass into breast milk as morphine at concentrations high enough to harm nursing infants. Several infant deaths have been linked to exactly this mechanism.

In 2013, the US Food and Drug Administration issued a black box warning — its most serious category — contraindicating codeine in children after tonsillectomy and adenoidectomy, following infant deaths in ultra-rapid metabolizer families. In 2017, the European Medicines Agency extended restrictions further. In 2024, the NHS completed the rollout of routine CYP2D6 testing before prescribing specific analgesics in pediatric settings.

The same story, with different supporting characters, plays out across dozens of commonly prescribed medications. It is not a rare edge case. It is a systematic gap between how medicine is currently prescribed and how it could be.

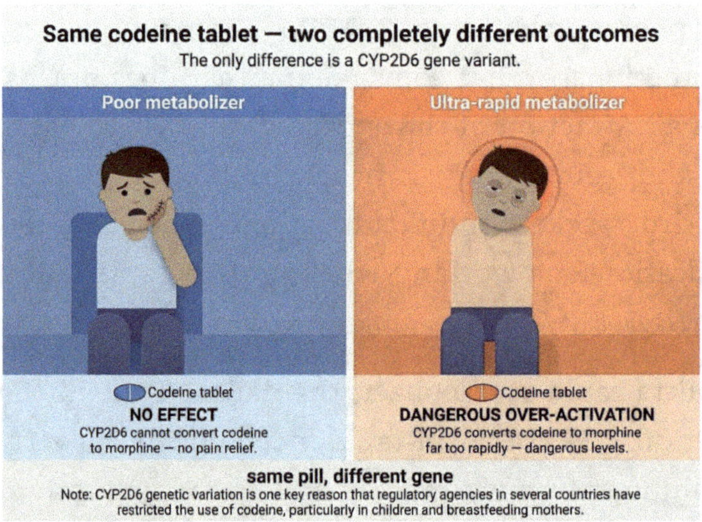

Figure 3: *Same codeine tablet, two completely different outcomes — left panel: poor CYP2D6 metabolizer, codeine not converted to morphine, no pain relief; right panel: ultra-rapid metabolizer, codeine converted to morphine far too rapidly, dangerous levels. Same pill, same dose, different gene.*
 Commissioned for The Biology of You series.

Every standard prescription is written for an average patient. Pharmacogenomics reveals whether you are that average patient — or whether a drug designed for someone else has just been handed to you.

Clopidogrel, Warfarin, and Your Heart: The Stakes When Blood Thinners Depend on Your Genes

If the codeine story illustrates pharmacogenomics in the context of pain management, the cardiovascular drug story illustrates it at a much higher level of clinical urgency.

Clopidogrel is one of the most widely prescribed medications in the world. It is an antiplatelet drug — it reduces the tendency of blood to form dangerous clots — and it is given to millions of people after heart attacks, strokes, and coronary stenting procedures to prevent recurrence. It is also a prodrug, which means it must be activated by the CYP2C19 enzyme before it has any effect at all.

Approximately 25 to 50 percent of the global population carries genetic variants that reduce or eliminate their CYP2C19 function. In these individuals, clopidogrel is not adequately activated in the body. They are taking a medication they have been told will protect their heart, and it is doing far less than either they or their doctor realizes. Studies have shown that CYP2C19 poor metabolizers on clopidogrel have significantly higher rates of heart attack and stroke recurrence than normal metabolizers on the same drug.

This is not a rare or theoretical problem. It is happening in a meaningful percentage of cardiology wards right now. In the UK, NICE issued guidance in 2024 recommending CYP2C19 testing to guide antiplatelet prescribing after stroke and transient ischaemic attack — a formal acknowledgement that genotype-guided prescribing in this area is evidence-based and clinically important. Alternative antiplatelet agents, such as prasugrel and ticagrelor, do not require CYP2C19 activation and may be more appropriate for poor metabolizers.

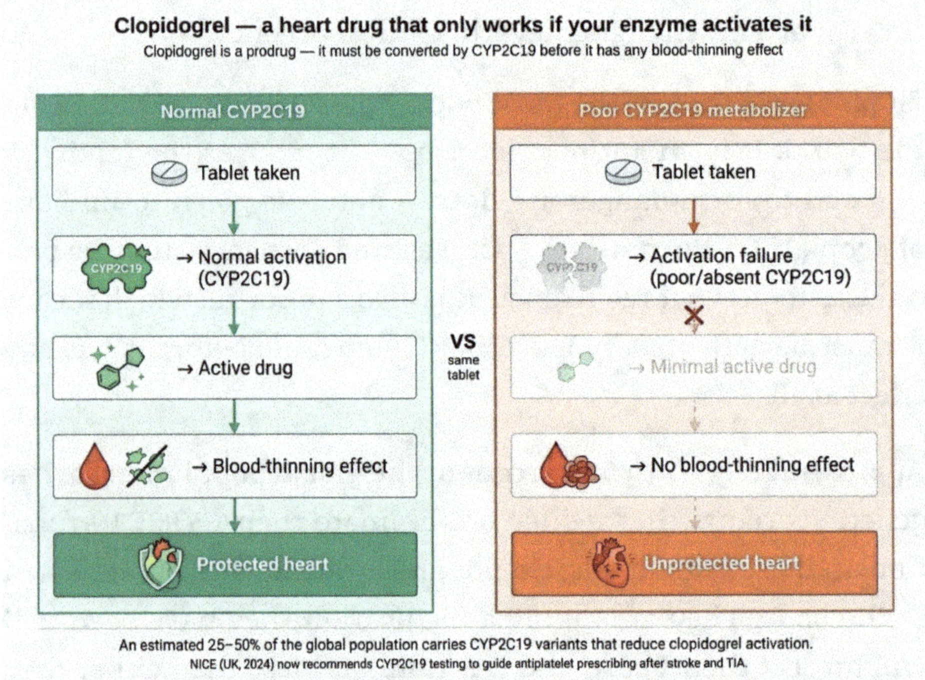

Figure 4: *Clopidogrel activation — two panels: normal CYP2C19 (tablet converts to active drug, platelets blocked, heart protected) versus poor CYP2C19 metabolizer (minimal conversion, platelets not blocked, heart unprotected). An estimated 25–50% of the global population carries CYP2C19 variants that reduce clopidogrel activation.*

Warfarin tells a complementary story. Warfarin is the most prescribed oral anticoagulant in history — a blood thinner used to prevent dangerous clots in people with atrial fibrillation, deep vein thrombosis, and mechanical heart valves. It is also the drug responsible for more adverse drug reaction hospitalizations in older adults than almost any other medication, because getting the dose right is genuinely difficult. Too little and clots form; too much and dangerous bleeding occurs.

Two genes explain a substantial portion of that variability. CYP2C9 determines how quickly warfarin is broken down; variants that reduce enzyme activity lead to warfarin accumulating to higher-than-expected levels. VKORC1 determines how sensitive the body's clotting mechanism is to warfarin's effects; variants in this gene explain why some people need five milligrams a day to achieve the same blood-thinning effect that one milligram produces in someone else. Together, variants in these two genes account for thirty to forty percent of the dose variability between patients. When a pharmacogenomic panel result is available before a warfarin prescription is initiated, studies show the time it takes to find a stable dose drops significantly, and the risk of serious bleeding events decreases.

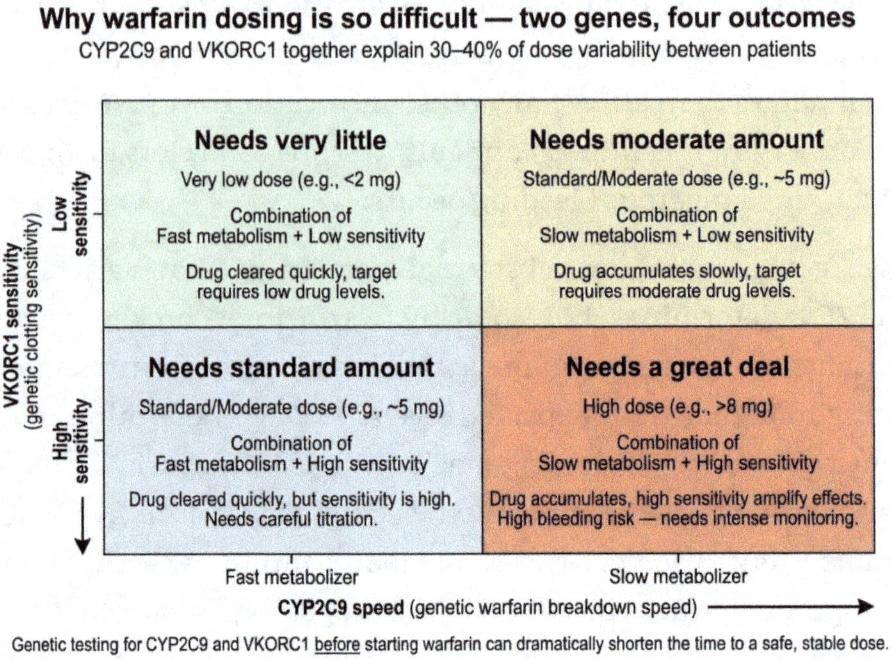

Figure 5: *Warfarin two-gene grid — CYP2C9 enzyme speed on one axis, VKORC1 clotting sensitivity on the other. Four quadrants show dose requirements from standard to very low. CYP2C9 and VKORC1 together explain 30–40% of dose variability between patients.*

Antidepressants and the Trial-and-Error Problem

There are perhaps no medications where the gap between the current reality of prescribing and the potential of pharmacogenomics is more personal, more consequential, or more avoidable than antidepressants.

The treatment of depression and anxiety often involves months of trial and error. A GP prescribes a first antidepressant. It either fails to work, or produces side effects so intolerable that the patient stops taking it, or both. A second is tried. Then perhaps a third. Each cycle takes weeks to play out, because antidepressants take time to reach therapeutic levels and time to assess. In the meantime, the patient is living with uncontrolled symptoms, accumulating experiences that antidepressants don't work for them, and sometimes abandoning treatment altogether.

CYP2D6 and CYP2C19 are the two enzymes responsible for metabolizing most commonly prescribed antidepressants, including the SSRIs fluoxetine, paroxetine, citalopram, and escitalopram, as well as many older tricyclic antidepressants. The relationship runs in both directions: CYP2D6 poor metabolizers experience higher blood levels of these drugs, increasing the risk of side effects like nausea, insomnia, and sexual dysfunction at doses considered normal. CYP2C19 poor metabolizers experience the same elevated drug levels with SSRIs metabolized through that pathway, potentially experiencing adverse effects that are mistaken for evidence that antidepressants simply don't suit them.

At the other end of the spectrum, CYP2D6 ultra-rapid metabolizers may clear antidepressants so quickly that a standard dose never builds to a therapeutically effective concentration. These patients often report that antidepressants have never worked for them — not because they have treatment-resistant depression, but because the drug is leaving their body faster than it can take effect.

A major randomized controlled trial — the PREPARE study, published in 2023 — examined the impact of pre-emptive pharmacogenomic testing across multiple drug classes including antidepressants. It found that genotype-guided prescribing reduced the incidence of clinically significant adverse drug reactions by 30 percent compared to standard prescribing. Not a marginal improvement. Thirty percent.

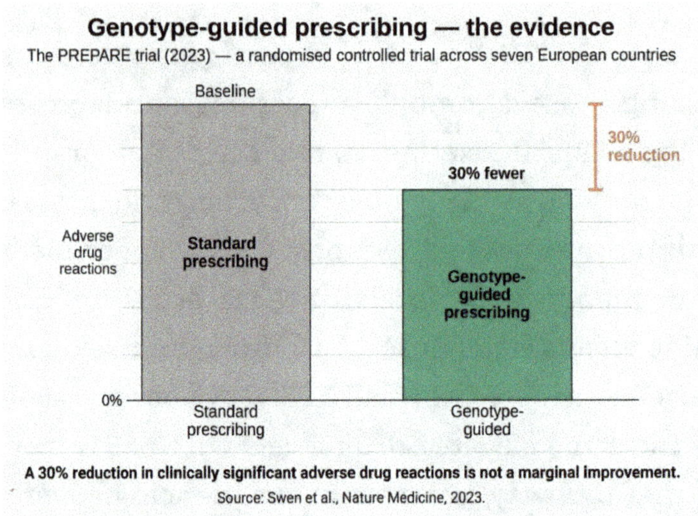

Figure 6: *The PREPARE study result — genotype-guided prescribing reduced clinically significant adverse drug reactions by 30% compared to standard prescribing across seven European countries.*
Source: Swen et al., Nature Medicine, 2023.
Bar heights are schematic — not precise figures.

Pharmacogenomic testing for antidepressant prescribing is available in the UK through companies such as Genomics England and through private services including GeneSight and Psomagen. Several NHS trusts are running pilots incorporating it into psychiatry pathways. In the US, GeneSight's Psychotropic test is the most widely used, and it has FDA clearance for the CYP2D6 and CYP2C19 components. The evidence is now clear enough that the question is no longer whether this testing is useful — it is how quickly it will become standard of care.

Pharmacogenomics doesn't make prescribing perfect. It removes one of the largest sources of unpredictability — and in mental health treatment, that unpredictability costs people months they cannot get back.

Chemotherapy and the DPYD Gene: When the Stakes Are Highest

If there is a category of pharmacogenomics where the case for routine testing is utterly unambiguous, it is in chemotherapy — and specifically in fluoropyrimidine chemotherapy, which includes 5-fluorouracil and capecitabine, among the most commonly used cancer drugs in the world.

These drugs are broken down in the body by an enzyme encoded by the DPYD gene. In most people, DPYD breaks down fluoropyrimidines efficiently, allowing the drug to work at its intended dose. But approximately 3 to 8 percent of people of European descent carry DPYD variants that significantly reduce this enzyme's function. In these individuals, the drug is not cleared properly. Concentrations build to levels that are not just toxic but can be fatal: severe mucositis, diarrhea, bone marrow failure, and death have been documented in DPYD-deficient patients given standard chemotherapy doses.

The solution is straightforward and well-established: test for DPYD variants before prescribing and reduce the dose accordingly in carriers. Studies show that dose-adjusted fluoropyrimidine treatment in DPYD intermediate metabolizers produces equivalent cancer outcomes with dramatically reduced toxicity. The test costs a few hundred pounds or dollars. The clinical benefit is unambiguous.

Pre-treatment DPYD testing has now been adopted as clinical policy in several countries, marking a clear shift in how the field views pharmacogenomics — from research interest to clinical necessity. In the United Kingdom, the NHS (National Health Service) began rolling out routine DPYD testing for all patients about to receive fluoropyrimidine chemotherapy as part of its NHS Genomic Medicine Service from 2020 onwards. In the United States, pre-treatment DPYD testing is recommended by the National Comprehensive Cancer Network (NCCN), the American Society of Clinical Oncology (ASCO), and the Clinical Pharmacogenomics Implementation Consortium (CPIC) — though implementation across oncology centers remains uneven and testing is not yet universally routine. If you or a family member is about to start fluoropyrimidine chemotherapy, ask your oncologist directly whether DPYD testing has been done. It should be standard. If it has not been offered, request it.

The same principle applies to thiopurine medications — azathioprine and 6-mercaptopurine — used in inflammatory bowel disease, autoimmune conditions, and certain cancers. TPMT and NUDT15 gene variants predict severe bone marrow toxicity in individuals who cannot adequately metabolize these drugs. Pretreatment testing is recommended by CPIC, by the American Gastroenterological Association, and in the UK by NICE, and is already routine in many clinical centers on both sides of the

Atlantic. If you are being prescribed azathioprine or 6-mercaptopurine, ask whether your TPMT and NUDT15 status has been checked.

When to ask your doctor about pharmacogenomic testing

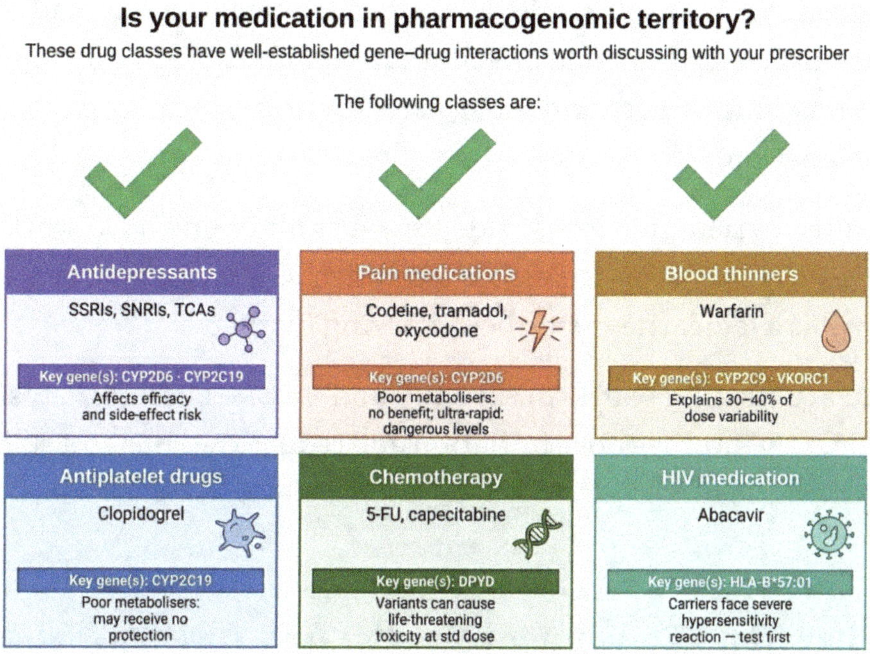

Figure 7: *Medication categories in pharmacogenomic territory — six tiles: antidepressants (CYP2D6 / CYP2C19), pain medications (CYP2D6), warfarin (CYP2C9 / VKORC1), clopidogrel (CYP2C19), fluoropyrimidine chemotherapy (DPYD), abacavir (HLA-B*57:01). If you take any of these, ask your prescriber whether pharmacogenomic testing has been considered.*

You have been prescribed a medication that requires CYP enzyme metabolism, and your doctor has not mentioned genetic testing. This is worth raising, especially for antidepressants, warfarin, clopidogrel, and any opioid analgesic.

You are about to start fluoropyrimidine chemotherapy or thiopurine therapy. Pre-treatment DPYD or TPMT/NUDT15 testing should be standard — ask specifically whether it has been done. In the US, it is recommended by NCCN, ASCO, and CPIC. In the UK, it is NHS policy. If your center has not offered it, request it before treatment begins.

A medication hasn't worked despite adequate doses and adherence or has caused side effects that seem out of proportion to the dose. Either scenario is consistent with a pharmacogenomic explanation.

You have tried multiple antidepressants and none have worked or all have caused intolerable side effects. A CYP2D6/CYP2C19 panel is a logical next step before trying further.

You are starting warfarin. CYP2C9 and VKORC1 results can significantly improve dose initiation and reduce the time to a stable therapeutic range.

What Your Consumer Genetic Test Already Tells You — and What It Doesn't

If you have already taken a consumer genetic test, you may have more pharmacogenomic information available than you realize.

23andMe's health reports include a pharmacogenomics section covering several clinically relevant genes, including CYP2D6, CYP2C19, and a handful of others, with FDA authorization for their drug response reports. The reports are clearly presented and include context about what each finding means for specific medications.

However — and this is important — consumer tests screen for a limited number of known variants in each pharmacogene. CYP2D6 in particular is an extraordinarily complex gene, with over a hundred identified variants and numerous structural variations including gene duplications and deletions that a standard SNP array cannot fully detect. A consumer test that reports you as a 'normal metabolizer' for CYP2D6 is providing a best estimate based on the variants it tested, not a complete analysis. Clinical pharmacogenomic testing uses more comprehensive methods, including specific copy number variation analysis, to characterize CYP2D6 fully.

For most common variants in most genes, consumer test results are accurate and clinically meaningful. For genes with significant structural complexity — particularly CYP2D6 — a clinical pharmacogenomics test offers more complete information and is worth requesting through your GP or a private testing service if the stakes are high.

Dedicated pharmacogenomics testing companies include GeneSight (strong in psychiatric medications, available in the US and UK), Genomics England (NHS-linked in the UK), OneOme RightMed (comprehensive panel, available through clinicians), and several others. These tests typically analyze between 27 and 120 variants across the most clinically relevant pharmacogenes and produce a report structured around specific drug classes — antidepressants, cardiovascular medications, pain management, oncology — making them directly actionable in conversation with a prescribing clinician.

The Drug Reference: Genes, Enzymes, and What They Mean for Your Prescriptions

The table below is a reference guide to the most clinically significant pharmacogenes covered by consumer and clinical testing. It is not exhaustive — the FDA's database of pharmacogenomic biomarkers in drug labeling currently lists over a hundred drugs — but it covers the genes and medications most likely to be relevant to your own healthcare.

Gene	Enzyme / Function	Key Medications Affected	Clinical Significance
CYP2D6	Metabolizes ~25% of all prescribed drugs	Codeine, tramadol, oxycodone, many antidepressants (fluoxetine, paroxetine, nortriptyline), antipsychotics (haloperidol, risperidone), tamoxifen	Poor metabolizers: drugs accumulate — side effects, toxicity, or no effect. Ultra-rapid metabolizers: drugs cleared too fast, or prodrugs over-activate (codeine toxicity risk). Frequency varies widely by ancestry: ultra-rapid metabolism affects ~1-2% of Europeans but up to 29% of East African populations.
CYP2C19	Metabolizes antiplatelet drugs, antidepressants, proton pump inhibitors	Clopidogrel (Plavix), omeprazole, escitalopram, citalopram, sertraline, diazepam	Poor metabolizers: clopidogrel not activated — elevated stroke/heart attack risk if used after cardiac event. ~12% of East Asian patients are poor metabolizers vs ~2% of Europeans. CPIC and AHA recommend testing before clopidogrel after stroke or TIA.
CYP2C9	Metabolizes warfarin, NSAIDs, and several other drugs	Warfarin, celecoxib, ibuprofen (high dose), phenytoin, losartan	Slow metabolizers require lower warfarin doses — standard doses risk bleeding. Combined with VKORC1, explains 30-40% of dose variability. Certain reduced-function variants (CYP2C9*5, *6, *8, *11) are more common in African-ancestry populations.
VKORC1	Target of warfarin — controls clotting factor synthesis	Warfarin	Variants alter sensitivity to warfarin. High-sensitivity variant requires lower doses; low-sensitivity variant requires higher doses. Acts

			together with CYP2C9 to determine individual warfarin dose requirements.
DPYD	Breaks down fluoropyrimidine chemotherapy drugs	5-fluorouracil (5-FU), capecitabine, tegafur	Poor metabolizers face life-threatening toxicity (severe mucositis, bone marrow failure) at standard doses. Pretreatment testing recommended by NCCN, ASCO, and CPIC in the US; NHS policy in the UK since 2020.
TPMT / NUDT15	Metabolize thiopurine immunosuppressants	Azathioprine, 6-mercaptopurine, thioguanine	Variants predict severe bone marrow toxicity. NUDT15 variants more common in East Asian and Hispanic populations; TPMT variants more common in European populations. Testing recommended by CPIC and the American Gastroenterological Association.
HLA-B*57:01	Immune system gene — controls drug hypersensitivity	Abacavir (HIV medication)	Carriers face a severe, potentially fatal hypersensitivity reaction. Universal pre-treatment testing recommended and widely implemented globally. Near-absent in European populations; present in ~6-8% of Han Chinese.
HLA-B*15:02	Immune system gene — controls drug hypersensitivity	Carbamazepine, phenytoin (epilepsy, neuropathic pain)	Carriers face Stevens-Johnson syndrome — a severe, life-threatening skin reaction. Near-absent in European populations; present in ~6-8% of Han Chinese and other Southeast Asian populations. Testing recommended before prescribing in Asian patients.
SLCO1B1	Transporter — controls statin uptake into liver cells	Simvastatin, atorvastatin, rosuvastatin	Reduced-function variants raise statin blood levels and increase myopathy (muscle damage) risk, most significantly with high-dose simvastatin. CPIC guidelines recommend dose adjustment or alternative statins for affected variants.
G6PD	Enzyme protecting red blood cells from oxidative damage	Primaquine, rasburicase, certain antibiotics and antimalarials	Deficiency causes hemolytic anemia when exposed to certain drugs. More common in populations from malaria-endemic regions — up to 20% in some African, Middle Eastern, and South Asian populations. Screening before primaquine

			for malaria treatment is standard.
UGT1A1	Metabolizes irinotecan chemotherapy	Irinotecan (colorectal and other cancers)	Reduced-function variant (*28 allele) impairs drug clearance, causing severe diarrhea and low white blood cell counts at standard doses. FDA labeling recommends dose reduction in homozygous *28 carriers. More common in African-ancestry populations.
CYP3A4 / CYP3A5	Metabolize approximately 50% of all prescribed drugs	Tacrolimus (organ transplant), cyclosporine, statins, many others	CYP3A5 expressers (more common in African-ancestry patients) metabolize tacrolimus faster and typically require higher doses. Widely tested in transplant medicine. CYP3A4 variants can affect a very broad range of medications.

The table is worth reading with your own medication list in hand. If you are currently prescribed any drug in the 'medications affected' column, it is worth having a conversation with your prescribing doctor about whether pharmacogenomic testing has been considered — and if not, whether it should be.

Population Differences: Why Your Ancestry Matters Here Too

Pharmacogenomics is the area of consumer genomics where the science applies most consistently across all human ancestries — and, strikingly, where the need is often greatest for people of non-European heritage. The reason is straightforward: the rates of non-normal drug metabolism are substantially higher in several non-European populations than in the European populations on which standard prescribing guidelines were built. This is not a limitation of pharmacogenomics. It is precisely the argument for why everyone — regardless of ancestry — should know their metabolizer status.

CYP2D6 ultra-rapid metabolism, caused by gene duplication, affects roughly 1 to 2 percent of people of Northern European descent. Among people of North African and Ethiopian heritage, that figure rises to 10 to 29 percent. The practical implication: an ultra-rapid metabolizer of North African heritage prescribed a standard antidepressant dose may clear the drug so quickly that it barely achieves any therapeutic effect and may be incorrectly considered treatment-resistant when the issue is metabolic, not medical.

CYP2C19 poor metabolizer status varies significantly across East Asian populations, where specific variants are far more prevalent than in European populations. CYP2C9 variants affecting warfarin metabolism are more common in some South Asian ancestry groups. HLA variants linked to drug hypersensitivity reactions vary in clinically significant ways by ancestry — the HLA-B*15:02 variant that predicts severe reactions to carbamazepine is almost entirely absent in European populations but present in approximately 6 to 8 percent of the Han Chinese population.

These differences carry an important implication. Clinical guidelines and standard drug dosing have historically been developed predominantly in studies of people of European descent — which means they were calibrated for the metabolizer frequencies seen in European populations. For anyone whose heritage falls outside those reference populations, standard prescribing may be even further from optimal than it would be for a European patient. The pharmacogenomic case for testing is, if anything, stronger. The caveat is that some consumer testing panels were built primarily on European variant catalogues and may not detect every clinically relevant variant for all ancestries — particularly certain CYP2D6 variants that are more common or unique to African-ancestry populations. For non-European patients who want the most complete picture, a clinical pharmacogenomic panel is more likely to detect ancestry-specific variants than a consumer test alone.

This is not a reason to avoid testing — it is a reason to test. The clinically actionable variants for the most important pharmacogenes are well-characterized across all major ancestry groups. For most people, a consumer pharmacogenomic result will be accurate and clinically useful regardless of heritage. Ensure your ancestry is noted when results are interpreted and raise it with any clinician using pharmacogenomic results to guide your treatment. If you are of non-European heritage and have been cautious about consumer genetic testing based on the ancestry bias discussed in Chapter Three, this chapter represents genuinely different territory: the findings here may matter more.

The Conversation to Have With Your Doctor

Pharmacogenomics is one of the areas where the gap between what the science can offer and what currently happens in a standard GP appointment is most apparent. Most doctors receive limited training in pharmacogenomics, testing is not yet universally available or funded, and the clinical infrastructure to integrate genetic results into prescribing decisions varies enormously between healthcare systems and between individual practices.

This does not mean you are without agency. It means you may need to be the person who introduces the conversation — and knowing how to do that is itself a form of health literacy worth having.

The most productive approach is to come to an appointment with something specific. Not 'I read about pharmacogenomics and think I should be tested,' but 'I've been taking this anti-depressant for six weeks with no improvement and I read that CYP2C19 variants can affect how people process this drug — is that something worth testing for?' Or: 'I'm about to start warfarin and I understand there are genetic variants that affect dosing — has that been considered?'

A doctor who is not familiar with pharmacogenomics is not a bad doctor. They are a doctor practicing in a healthcare system that is still in the early stages of integrating this science into routine care. The question you are asking is legitimate, evidence-based, and increasingly likely to be met with interest rather than dismissal.

Increasingly, pharmacogenomic testing is also available directly through private laboratories and testing companies without a GP referral. If your healthcare system does not yet offer it routinely, this route is worth exploring — particularly for anti-depressants, where the evidence for clinical utility is strong and the cost of trial-and-error prescribing, measured in months of suffering, is high.

You are not asking for special treatment. You are asking for prescribing that reflects who you actually are — not who the average patient in a clinical trial was assumed to be.

What We Covered in This Chapter

Here are the ideas to carry forward:

- Pharmacogenomics is the study of how your genetic profile shapes your response to medications — whether a drug works, how well it works, and whether it causes you harm.
- CYP enzymes in your liver metabolize most drugs. The genes that build these enzymes come in variants that make you a poor, intermediate, normal, or ultra-rapid metabolizer — with significant consequences for drug efficacy and safety.
- Codeine, clopidogrel, warfarin, antidepressants, and fluoropyrimidine chemotherapy drugs are all areas where pharmacogenomic variants have well-established, clinically actionable implications. Knowing your status before prescribing can improve outcomes and prevent serious harm.
- The PREPARE study found that genotype-guided prescribing reduced clinically significant adverse drug reactions by 30 percent. Pre-treatment DPYD testing for fluoropyrimidine chemotherapy is now NHS policy in the UK and recommended by NCCN, ASCO, and CPIC in the US. CYP2C19 testing before clopidogrel prescribing after stroke is now recommended by NICE in the UK and by the American Heart Association and CPIC in the US. The field is moving fast — clinical adoption is following the science.

- Consumer genetic tests provide useful pharmacogenomic information but may not fully characterize complex genes like CYP2D6. Clinical pharmacogenomics panels offer more complete analysis, especially where the stakes are high.
- Pharmacogenomics is the most ancestry-inclusive field in consumer genomics: the mechanism is biochemical and universal, and the rates of non-normal drug metabolism are often higher in non-European populations than in the European populations on which standard prescribing was calibrated. If your heritage is non-European, pharmacogenomic testing may be more important for you, not less — though a clinical panel will give the most complete picture of ancestry-specific variants.
- You can and should raise pharmacogenomics with your prescribing doctor — especially when medications have not worked, have caused disproportionate side effects, or when you are about to start a drug in a high-risk category.

Coming up in Chapter Five — Whole Genome Sequencing: The Full Picture. What it means to have every letter of your genome read, what is currently interpretable and what isn't, who should consider it, where to access it, and what to expect when you receive the most complete genetic information a test can provide.

✦ ✦ ✦

Chapter Five

Whole Genome Sequencing

The Full Picture — What It Is, What It Costs, and What You Actually Get

In 2003, the most ambitious biological project in human history reached its conclusion. After thirteen years of work, a budget of around three billion dollars, and the coordinated effort of scientists across twenty universities and research institutes in six countries, the Human Genome Project delivered its final product: the first complete reading of a human genome.

Every one of the roughly three billion letters in the human instruction set — read, for the first time, in their entirety.

It was a landmark moment. And it was, almost immediately, a benchmark that the following two decades of scientific progress set about demolishing. The cost of sequencing a human genome has fallen faster than almost any technology in history — faster than computer processing power, faster than hard drive storage, faster than solar panels. The same sequence that cost three billion dollars in 2003 can be read today for somewhere between two hundred and six hundred dollars, depending on the service and the depth of analysis you choose.

This chapter is about what that means for you — as someone who can now, if they choose, have every letter of their genome read. Not a curated selection of half a million positions, as a consumer SNP array provides. Not a targeted panel of genes known to be medically relevant. Every letter. All of it. The most complete biological self-portrait a person has ever been able to hold.

And, just as importantly, this chapter is honest about what that completeness actually gives you — and what it does not. Because the sequencing itself is, today, the easy part. What the world is still working out is how to interpret what it finds.

We can now read every letter of your genome faster than you could read this book. The harder question — and the one science is still answering — is which letters, out of three billion, are actually telling you something you need to know.

What Whole Genome Sequencing Actually Is

The consumer tests we explored in Chapter Two — the 23andMe kits, the AncestryDNA swabs — use a technology called SNP array genotyping. As we established, an SNP array checks specific known positions in your genome — usually around 650,000 to 900,000 of them — and tells you which of two possible letters you carry at each position. It is targeted, efficient, and relatively cheap. But it is checking less than 0.03 percent of your total genome.

There is a dimension to this distinction that matters particularly if your ancestry is not primarily European. The specific positions that SNP arrays check were not chosen at random — they were selected because they appeared frequently in genome-wide association studies, the overwhelming majority of which were conducted in people of European descent.

A SNP array therefore checks a set of positions that are informative for European ancestry genomes, but which may not capture the most relevant variants in your genome if your heritage is from Africa, East Asia, South Asia, the Middle East, or Latin America.

Whole genome sequencing, by reading every position regardless of prior knowledge, does not have this starting-point bias. This is one reason why WGS is, in practice, the more ancestry-equitable of the two technologies — even though the interpretation of what it finds still reflects existing database limitations.

Whole genome sequencing is different in nature, not just in scale. Rather than checking pre-selected positions, it reads every base pair across your entire genome — all three billion of them. It does this by breaking your DNA into millions of short fragments, reading each fragment independently, and then using sophisticated software to reassemble the readings into a complete, coherent sequence — like putting back together a document that has been shredded into millions of tiny pieces, each piece read simultaneously.

The depth at which this reading is done matters enormously, and it is described in terms of coverage — how many times, on average, each position in the genome has been read independently. Low-pass whole genome sequencing might read each position as few as four times. Standard consumer whole genome sequencing uses 30× coverage — each position read thirty times — which is considered the clinical gold standard for detecting variants with high confidence. Anything below 15× starts to introduce meaningful error rates; anything above 30× is typically used only for specific clinical applications such as cancer genomics, where rare mutations in a small subset of cells need to be detected.

When companies advertise 30× whole genome sequencing, this is the number to look for. It means that the sequence you receive has been generated with sufficient depth and redundancy to be reliable for health interpretation — though as we will explore shortly, reliability of the sequencing and reliability of the interpretation are two very different things.

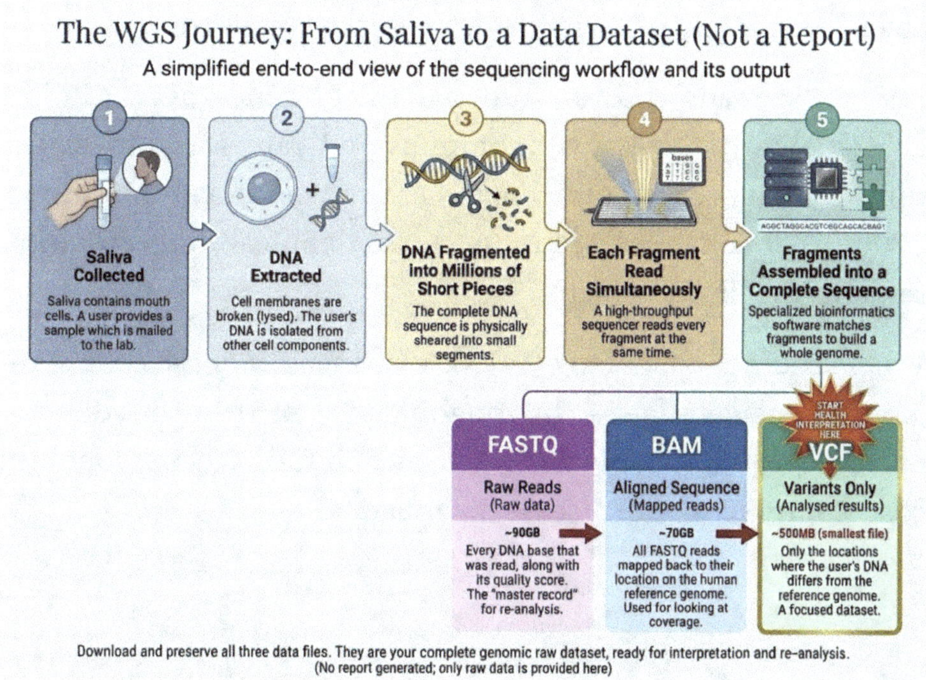

Figure 1: *The whole genome sequencing process — five steps from saliva to three output file types: FASTQ (~80-200 GB raw reads), BAM (~60-100 GB aligned sequence), and VCF (~500 MB variants only).*

When companies advertise 30× WGS, each position in your genome has been read independently thirty times, giving high confidence in the result.

SNP array vs. whole genome sequencing — at a glance

SNP array: checks 650,000–900,000 known variant positions. Covers less than 0.03% of the genome. Fast, affordable ($80–$200). Excellent for ancestry, disease risk scores, pharmacogenomics, and carrier status. Cannot detect rare variants outside the tested positions, structural changes, or novel mutations.

Low-pass WGS: reads the full genome at 4–8× coverage. Better than SNP array for rare variant detection but not clinically reliable. Used mainly in research and ancestry applications.

30× WGS: reads the full genome thirty times over. Clinical gold standard. Detects all common and rare variants, structural changes, copy number variations, and novel mutations. $200–$600 consumer; $1,000–$3,000+ clinical. Interpretation remains the limiting factor.

Whole exome sequencing (WES): sequences only the coding regions of the genome — the roughly 1–2% of DNA that directly encodes proteins. Between SNP array and full WGS in scope and cost. Common in rare disease diagnostics.

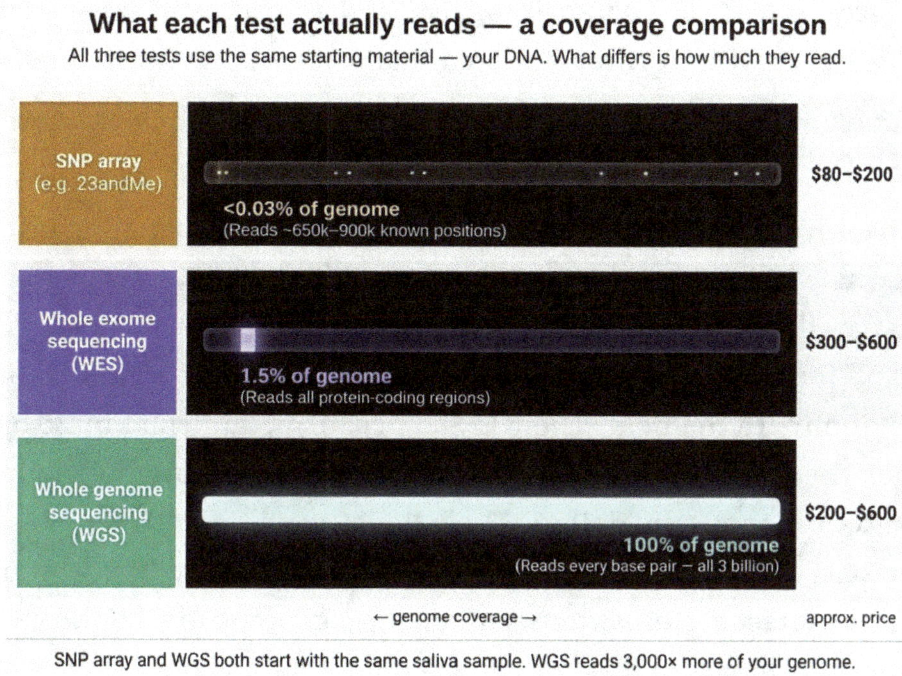

Figure 2: *Coverage comparison — three horizontal bars showing the proportion of the genome read by SNP array (<0.03%), whole exome sequencing (~1-2%), and whole genome sequencing (100%).*
Bar widths are illustrative — the true scale difference cannot be shown proportionally.
SNP array positions were selected predominantly from European GWAS studies.

The Price Revolution: From Three Billion Dollars to Under Three Hundred

The story of how genome sequencing became affordable is one of the most remarkable examples of technological cost deflation ever recorded — and understanding it helps you make sense of why the consumer market looks the way it does.

The original Human Genome Project used a technique called Sanger sequencing — the same method Francis Sanger developed in the 1970s — scaled up to an almost incomprehensible degree. It was slow, labor-intensive, and enormously expensive. The reason it cost three billion dollars was not scientific inefficiency but the genuine difficulty of the problem at the scale being attempted.

The revolution came from next-generation sequencing platforms — machines that could read millions of DNA fragments in parallel rather than one at a time. Illumina, a San Diego company founded in 1998, built the instruments that would come to dominate the field. Their HiSeq platform, launched in 2010, was the first to bring the cost of sequencing a human genome below ten thousand dollars. Their NovaSeq platform, introduced in 2017, brought it below a thousand. By the early 2020s, Illumina's latest systems were sequencing a genome for around two hundred dollars in reagent costs at commercial scale.

This is the technology that powers the consumer sequencing services available today. When Nebula Genomics offers 30× whole genome sequencing for $299, or Dante Labs for $399 to $999 depending on the reporting package, they are running your sample through Illumina or equivalent next-generation sequencers operating at enormous volume, passing the economies of scale directly to the consumer. The sequencing itself is genuinely affordable. What you are partly paying for beyond the raw sequencing cost is the software, the reports, the data storage, and the company's ongoing ability to interpret and update your results as science advances.

The next chapter in this cost story is already being written. Long-read sequencing platforms — particularly those developed by Oxford Nanopore Technologies and Pacific Biosciences — can read much longer fragments of DNA in a single pass, which dramatically improves the detection of structural variants and repetitive regions of the genome that short-read Illumina technology misses. These platforms are already in use in clinical settings and are beginning to appear in consumer offerings. As their cost continues to fall, a 30× short-read sequence may within a few years be considered the baseline rather than the gold standard.

The sequencing bottleneck is gone. A technology that once required a fifteen-year international project now fits in a courier envelope and returns results in a few weeks. What has not yet caught up is our ability to tell you what all of it means.

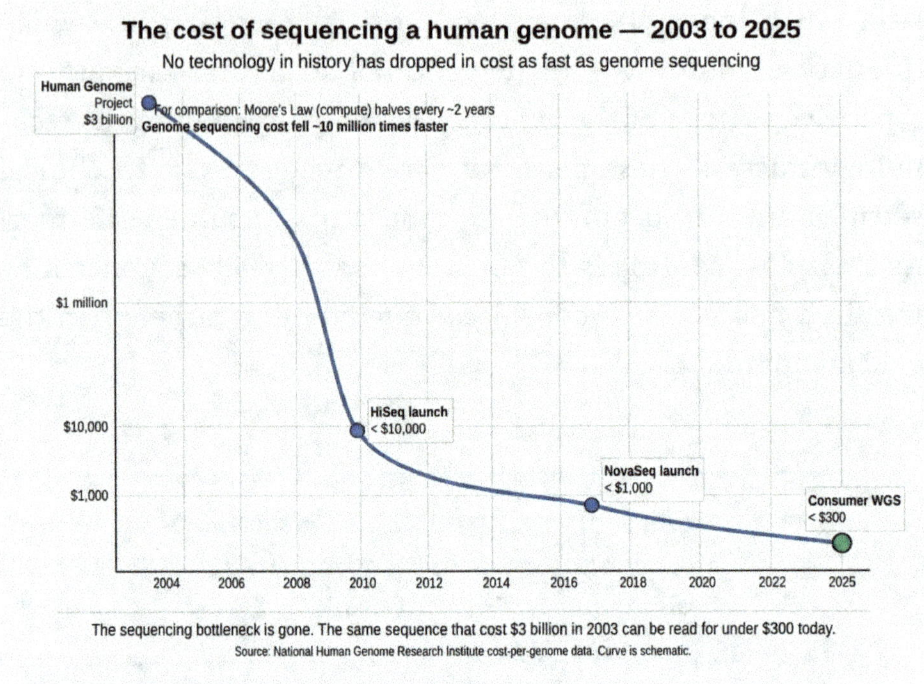

Figure 3: *A logarithmic cost curve running from 2003 to 2025, with labelled milestones (Human Genome Project 2003: $3 billion → HiSeq 2010: under $10,000 → NovaSeq 2017: under $1,000 → consumer services 2025: under $300), communicates not just the price drop but the acceleration of that drop. The shape of the curve — steep, then steeper — is the visual argument that this is not normal technological progress. It is something historically unprecedented.*

The Interpretation Gap: What We Can and Cannot Yet Read

Here is the most important thing to understand about whole genome sequencing in 2025, and the thing that consumer marketing tends most consistently to obscure: sequencing your genome and understanding your genome are not the same thing.

Of the roughly three billion base pairs in your genome, science has established the function and significance of a relatively small proportion. The protein-coding regions — the roughly 2 percent of the genome that directly encodes genes — are relatively well characterized. The clinical significance of many individual variants in established disease-associated genes is known and actionable. But the remaining 98 percent of the genome, once dismissed as junk DNA, is now understood to play important regulatory roles — and the science of interpreting what those roles mean for individual health is still in its early stages.

When a 30× whole genome sequence is generated and analyzed, the reporting software identifies your variants by comparing your sequence to a reference genome. A variant in a well-characterized gene — say, a pathogenic variant in BRCA2 — will be flagged as clinically significant. A variant in a less-studied region of the genome will be classified as a variant of uncertain significance, or VUS. And the honest reality is that the majority of variants a whole genome sequence identifies fall into this uncertain category — not because they are necessarily harmless, but because science simply has not yet studied them in enough people to establish what they mean.

WGS and ancestral diversity — what non-European ancestry readers should know

When your WGS results are generated, your variants are identified by comparing your sequence to a reference genome — a composite sequence that serves as a baseline for human DNA. The current standard reference (GRCh38) was assembled primarily from a small number of donors, with one individual of African-European origin accounting for approximately 70% of the sequence. This means the reference genome itself underrepresents the genetic diversity present in non-European populations.

> The practical consequence: non-European ancestry individuals typically receive approximately 23% more variants of uncertain significance (VUS) than European ancestry individuals receiving the same test — not because their genomes are more uncertain, but because the reference databases have fewer entries for their population-specific variants. This is a limitation of the current databases, not of your biology.
>
> This makes the re-analysis argument from the next paragraph especially important for non-European ancestry individuals. The databases are diversifying faster than at any previous point — driven by the NIH All of Us program (750,000+ diverse participants), the African Ancestry Genome Project, and the GenomeAsia 100K project. Many variants classified as VUS today will be reclassified as the data grows. A sequence taken now and stored carefully is a resource that will become more informative over time — and the rate of improvement is particularly high for non-European ancestry genomes.

> The Human Pangenome Reference Consortium is developing a new reference that incorporates 47 haplotype-diverse genomes from across the world. When this becomes the clinical standard, VUS rates for non-European ancestry individuals will fall significantly. Your sequence will be re-interpretable against it.

This does not make whole genome sequencing useless. It makes it an investment in a resource that will increase in value over time. A sequence taken today and stored carefully can be re-analyzed in five years, ten years, twenty years — as the databases of variant significance grow, as the populations studied become more diverse, as the regulatory regions of the genome are gradually decoded. Many people who receive a VUS today will find, in a future re-analysis, that it has been reclassified — either as benign, which closes a question, or as pathogenic, which opens a clinical conversation that could not have been started without the sequence existing in the first place.

The analogy that serves well here is the early internet. In 1995, an internet connection gave you access to a vast amount of information — but the search tools to find what you actually needed were primitive, and most of what was available was either incomprehensible or not yet uploaded. The infrastructure existed before the knowledge base that made it fully useful. Whole genome sequencing is in a similar moment. The infrastructure — the sequencing technology, the storage, the bioinformatics pipelines — is mature. The knowledge base that will make every corner of that sequence interpretable is still being built, one published study at a time.

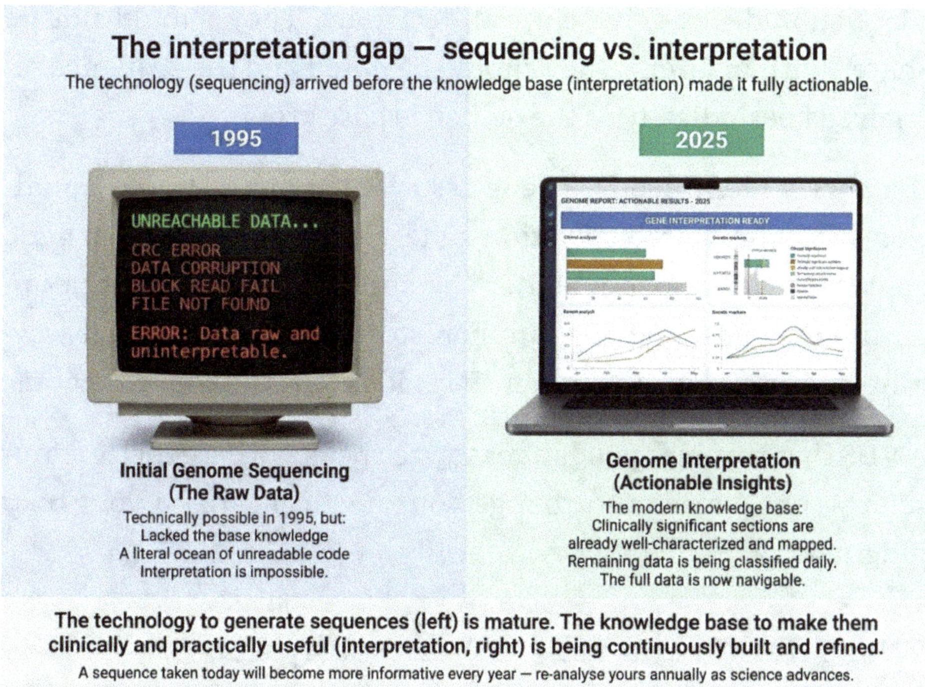

Figure 4: *The early internet analogy — left panel (1995): internet connected, vast information available, but primitive search tools and most content unreachable. Right panel (2025): same information, now navigable and useful. Applied to genome interpretation: the sequencing infrastructure is*

mature; the knowledge base is being built. The gap is narrowing faster for non-European ancestry populations than at any previous point.

What a variant of uncertain significance (VUS) actually means

A VUS is not a diagnosis. It is not a risk factor. It is a marker of current scientific uncertainty — a variant that has been detected but whose effects, if any, are not yet established.

The majority of people who receive whole genome sequencing will have VUSs in their report. This is expected and does not indicate a problem.

VUSs should not drive clinical decisions. They should not be shared with insurers. They should be noted and the sequence reanalyzed periodically as the evidence base grows.

If a VUS is identified in a gene associated with a serious condition — a cancer predisposition gene, a cardiac gene — a genetic counselor can assess whether the specific change, its location within the gene, and any family history context suggest it warrants monitoring, or whether the evidence is too thin to act on.

A VUS that troubles you is a reason to speak with a genetic counselor — not a reason to change your medical care on your own judgment.

Who Offers It and What You Actually Get

The consumer whole genome sequencing market has matured considerably since the first sub-thousand-dollar tests became available around 2017. The major players each make different trade-offs between price, reporting depth, data ownership, privacy, and clinical credibility. Here is an honest assessment of what is available.

Nebula Genomics, founded in 2018 as a Harvard Medical School spinout by George Church and his collaborators, was among the first companies to offer consumer 30× whole genome sequencing at an accessible price point — eventually reaching $249 to $299 for sequencing, with an ongoing subscription of around $19 per month for access to their report library. Their reporting suite covers over 300 traits and health areas, updated weekly as new research emerges. Their privacy model is notably strong — they give users direct control over data sharing and have built blockchain-based access controls into their platform. The sequencing is performed at large-scale facilities and the results are returned in standard genomic file formats that you can download and use independently of any ongoing subscription.

Dante Labs, an Italian company founded in 2016 with a New York office, offers 30× whole genome sequencing using Illumina NovaSeq technology with ISO 15189 medical laboratory accreditation — a meaningful distinction, as this is the certification required for clinical diagnostic laboratories. Their pricing ranges from roughly $399 to $999 depending on the reporting package, with more than 200 health reports across cardiology, oncology, neurology, pharmacogenomics, and rare disease. Their portal is designed to produce clinician-ready reports, making Dante's offering more naturally positioned for integration with medical care than most consumer tests. Historically, the company has received mixed reviews regarding customer service and turnaround times, though the quality of the underlying sequencing has generally been regarded as sound.

Sequencing.com takes a different approach: rather than positioning primarily as a sequencing provider, they function more as a genomic data platform. They offer 30× whole genome sequencing, but they also accept raw data files from other testing companies — meaning if you have already taken a 23andMe or AncestryDNA test, you can upload that data and run it through their interpretation apps. Their app ecosystem includes hundreds of reports from third-party developers, which creates both more variety and more variability in quality than a single-company report library.

For people who want clinical-grade sequencing ordered through a physician, Invitae offers whole genome sequencing as part of their clinical testing portfolio, with turnaround through a physician referral and results returned to both patient and clinician. Their testing is CLIA and CAP certified — the US standards for clinical laboratory quality — and their variant interpretation database is among the most extensively curated in the industry. This is the route most likely to result in findings that integrate smoothly into your medical care, at a correspondingly higher cost.

In the United Kingdom, the NHS Genomic Medicine Service offers whole genome sequencing at no cost to patients with a clinical indication — rare undiagnosed disease, cancer, and hereditary conditions with a suspected genetic basis. The 100,000 Genomes Project, completed in 2018, laid the infrastructure for what is now a routine NHS service, and by August 2024 the NHS had sequenced its 100,000th patient genome through the NHS GMS. The Generation Study, launched in 2023, is now sequencing the genomes of 100,000 newborn babies to explore the potential of genomic newborn screening for over 200 rare, treatable conditions. For UK residents without a clinical indication, the consumer services above are the current route to WGS outside the clinical pathway.

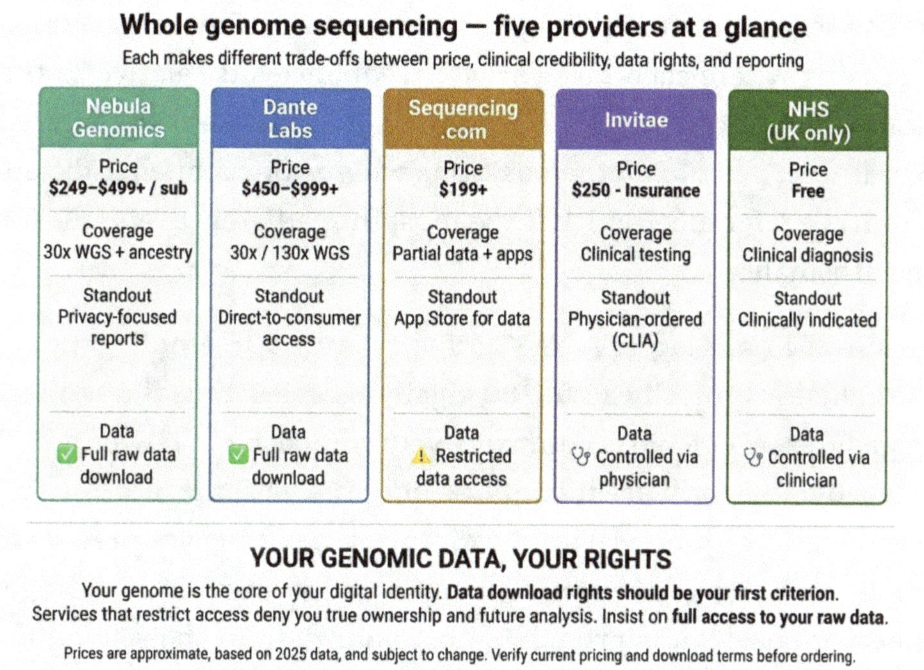

Figure 5: *Five whole genome sequencing providers at a glance — Nebula Genomics, Dante Labs, Sequencing.com, Invitae (clinical-grade), and NHS (UK, clinical indication). Each card shows price, coverage depth, standout feature, and data download rights. Data download rights should be the first criterion for any consumer.*

What You Actually Receive: The Files, the Reports, and the Data

When your whole genome sequencing results arrive, they come

in two distinct forms: raw data files and interpreted reports. Understanding both — and what to do with each — is the practical heart of this chapter.

The raw data files are the actual sequencing output. Three file types are standard. The FASTQ file contains the raw sequence reads — millions of short DNA fragments, each with an associated quality score. The BAM file is the aligned version — those fragments mapped against the reference human genome, showing where each piece of your sequence sits relative to the known map. The VCF file — variant call format — is the filtered summary: a list of all the positions where your sequence differs from the reference, which is the most immediately useful file for health analysis.

These files are large. A FASTQ file from a 30× whole genome sequence is typically 80 to 200 gigabytes. A BAM file is 60 to 100 gigabytes. A VCF file is comparatively compact — a few hundred megabytes — because it contains only the variants, not the full sequence. The practical implication: downloading your raw data requires a stable internet connection, significant local storage, and patience. Some providers host your data in their cloud indefinitely; others offer a download window of thirty days before charging for ongoing storage access. Reading your raw data files requires specialist bioinformatics software — they are not human-readable documents.

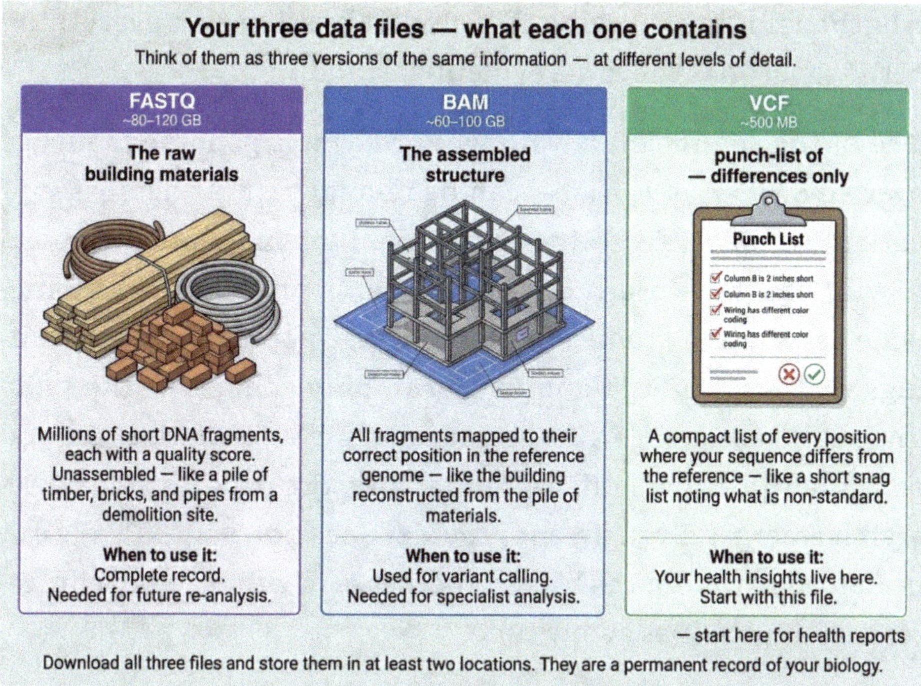

Figure 6: *Your three data files — FASTQ as raw building materials (~80-200 GB), BAM as the assembled structure (~60-100 GB), VCF as the punch-list of differences from the reference (~500 MB). The VCF is the most immediately useful starting point for health interpretation. Download and store all three files permanently.*

The interpreted reports are what most people actually engage with. These translate the variants identified in your VCF file into language about health risks, traits, pharmacogenomics, and ancestry. The quality of these reports varies significantly between providers. The best reports — from Dante Labs' clinical platform, from Invitae's physician-ordered service, from well-curated consumer portals — are based on peer-reviewed databases of variant significance, cite the evidence behind each finding, and clearly distinguish between well-established variants and variants of uncertain significance. Less rigorous reporting systems may present a VUS with the same visual weight as a clinically established finding, creating the false impression that everything in the report has the same evidential standing.

One of the significant advantages of a whole genome sequence over an SNP array is that it can be re-analyzed. As databases of variant significance grow — as ClinVar, the main public repository of clinical variant classifications, is updated; as genome-wide association studies in more diverse populations add findings that were invisible in European-only cohorts — the same underlying sequence file becomes more informative. Several consumer providers offer automatic report updates as new research emerges. Even if your provider does not, your VCF file can be uploaded to services like Promethease or run through clinical reanalysis pipelines at any time.

This is the reason that downloading and safely storing your raw data files is one of the most important practical steps following a whole genome sequence. The sequence you generate today is, in a meaningful sense, a resource for the rest of your life.

How to store and protect your genomic data

Download your VCF file immediately upon receiving results and store it on at least two independent devices — an external hard drive and a cloud backup service. Unlike photographs or documents, this file will not become obsolete: it can be re-analyzed indefinitely as interpretation improves.

If your provider allows FASTQ or BAM download, consider preserving these too if you have the storage capacity. These contain richer data than the VCF and may be necessary for certain future analyses.

Do not store your raw genomic data in the same cloud account where you store personal identifying documents. Keep it separate, or use a service with strong encryption and a clear data deletion policy.

Consider using a pseudonymous email address and separate account for genomic data services — particularly consumer services — as an additional layer of privacy protection.

If your provider is acquired, changes its privacy policy, or goes out of business — as happened with 23andMe in 2025 — you want your data in your own hands before that happens, not locked in their platform.

The 23andMe Lesson: Why Data Ownership Matters More Than You Think

In March 2025, 23andMe filed for bankruptcy. The company that had built one of the world's largest private databases of human genetic information — over fifteen million people's genomes — was acquired through a bankruptcy sale, raising immediate questions about what would happen to the genetic data of the people whose saliva was sitting in their archives.

The California Attorney General issued guidance urging 23andMe users to delete their data from the platform. Privacy advocates raised concerns about who the acquiring entity would be and what terms would govern their use of the existing database. The situation was, in the end, navigated more carefully than the worst fears suggested — but it served as a vivid illustration of a risk that most people who take genetic tests never consider: the company you trust with your genome today may not exist in its current form tomorrow.

The lesson is not that genetic testing is dangerous or that you should avoid it. The lesson is that the value of your genomic data is permanent and portable — but only if you treat it that way. A company can go bankrupt, be acquired, change its privacy policy, or be subject to a data breach. Your VCF file, downloaded and stored in your own hands, cannot be affected by any of those events.

This is the principle behind choosing providers who allow full data download, and why data portability should be considered alongside price and report quality when selecting a sequencing service. Of the major providers, Nebula Genomics, Dante Labs, and Sequencing.com all allow raw data download. Invitae returns your results through your physician. The NHS retains clinical records but provides variant reports. When evaluating any service, ask explicitly: can I download my complete raw data, at any time, at no additional cost, for as long as I choose?

Who Should Consider Whole Genome Sequencing Now?

Given everything this chapter has covered — the genuine power of a complete genome sequence, the current limits of interpretation, the data ownership considerations, and the cost landscape — the question becomes personal: is this the right step for you, and if so, when?

Whole genome sequencing is most clearly valuable in three situations. First, for people with a strong family history of a genetic condition that has not yet been identified at the variant level. If your family has a pattern of hereditary cancer, early-onset heart disease, a neurological condition, or unexplained deaths in younger relatives, and targeted genetic testing has not identified the causative variant, a whole genome sequence offers the most comprehensive single sweep of your genetic landscape. It is increasingly the test of choice in rare disease diagnosis — a New England Journal of Medicine study using the 100,000 Genomes Project data found that WGS delivered new diagnoses across the broadest range of rare diseases studied to date, in patients who had previously received no genetic explanation. Second, for people of non-European ancestry who want the most accurate picture of their genetic health risks. As established in Chapter Two and throughout this book, SNP arrays check positions preselected from predominantly European GWAS studies — meaning they may miss population-specific variants and return less accurate risk scores for non-European ancestry individuals. WGS reads everything, regardless of prior knowledge, and is the more ancestry-equitable choice. The VUS rate will be higher today, but re-analysis as databases diversify will be particularly rewarding. Third, for anyone thinking long-term. If you are in your thirties or forties, in reasonable health, and interested in having the most complete possible baseline of your genetic profile as a resource for future medical decisions, a 30× whole genome sequence is a reasonable investment. The sequence will not change. The science interpreting it will advance continuously. The combination means a sequence taken now becomes more valuable, not less, as the years pass.

It is also valuable for anyone who is thinking long-term. If you are in your thirties or forties, in reasonable health, and interested in having the most complete possible baseline of your genetic profile as a resource for future medical decisions, a 30× whole genome sequence is a reasonable investment. The sequence will not change. The science interpreting it will advance continuously. The combination of those two facts means that a sequence taken now becomes more valuable, not less, as the years pass.

It is less immediately useful for someone expecting a consumer-style health report with clear, actionable findings on every page. A SNP array test from 23andMe or AncestryDNA, despite covering a tiny fraction of the genome, is currently better at delivering clean, well-interpreted health risk scores for common conditions — simply because those scores are built from GWAS studies of the specific positions those tests check, and the reporting infrastructure for those findings is mature. A whole genome sequence offers more data but, today, not necessarily more actionable insight for the average consumer. The exception is if your ancestry is non-European: because the positions SNP arrays check were pre-selected from European GWAS studies, an SNP array gives non-European ancestry individuals a less accurate picture than it gives European ancestry individuals. For those readers, WGS is the more equitable starting point even if the interpretation currently has more gaps.

The honest framing is this: a whole genome sequence is the most complete version of a document that science is still learning to read fully. Taking one now is not premature — the clinically important sections are well enough understood to be genuinely useful. But it requires a tolerance for uncertainty, an appreciation that many of the findings will be marked as uncertain significance today and reclassified in the future, and a commitment to treating the raw data as a long-term asset rather than a one-time report.

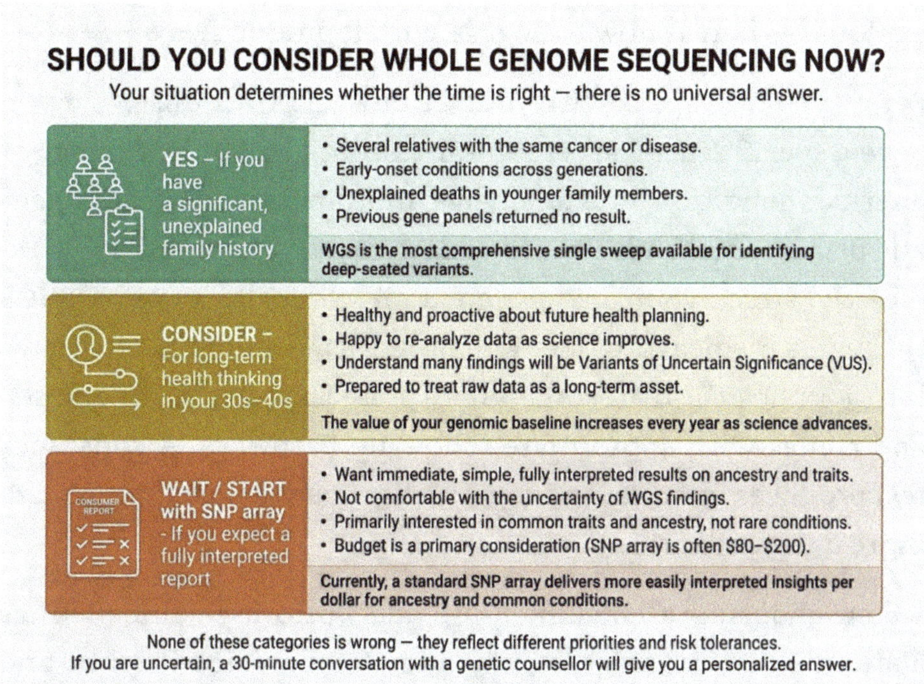

Figure 7: *Is whole genome sequencing right for you now? — three recommendation cards: Yes (unexplained family history, OR non-European ancestry seeking a more equitable test than SNP arrays), Consider (long-term thinking, 30s-40s), and Wait or start with SNP array (expecting a fully interpreted consumer report immediately).*

A whole genome sequence taken today is not just a snapshot of what science knows now. It is a resource that becomes more legible every year, as more people are sequenced, more variants are characterized, and the remaining 98 percent of our genome gradually gives up its secrets.

Getting the Most From a Whole Genome Test: A Practical Guide

If you decide to proceed with whole genome sequencing — or if you already have a sequence sitting in a portal you haven't fully explored — here is how to engage with it productively.

Before you test, speak with your GP or a genetic counselor if you have a specific medical reason for testing — a family history of a serious condition, unexplained symptoms with a possible genetic origin, or a condition for which genetic information would directly affect your management. A clinically ordered WGS through a medical provider will produce a result that integrates into your healthcare record, is interpreted by trained geneticists, and comes with appropriate follow-up pathways. A consumer test produces raw data and reports that you will need to navigate more independently.

When choosing a consumer provider, prioritize three things above all: 30× coverage, FASTQ and VCF download rights, and a transparent privacy policy. Everything else — the reporting interface, the ancestry features, the subscription model — is secondary to having a reliable sequence and ownership of your own data.

When your results arrive, start with the clinically established findings. Most platforms flag these separately from the general trait reports — variants in known disease genes with established clinical significance. These are the findings worth discussing with a doctor. For everything else, approach the reports with appropriate skepticism: a report on a complex trait like intelligence, personality, or exercise response, generated from a handful of variants, is interesting but far from deterministic.

Find out whether your provider updates reports automatically. Nebula Genomics publishes weekly research library updates. Dante Labs updates reports as part of their subscription model. If your provider does not automatically update, set a reminder to re-upload your VCF file to a reanalysis service — Promethease, for example — once a year. The science moves fast enough that an annual reanalysis is a worthwhile discipline.

Finally, tell your doctor you have a whole genome sequence. You do not need to hand over the entire raw file — but letting your GP or any specialist know that you have genomic data available means that, if a situation arises where genetic information would change a clinical decision, the conversation about accessing and interpreting it can happen quickly rather than starting from scratch.

A note on direct-to-consumer WGS and clinical validity

Consumer whole genome sequencing services are not the same as clinical diagnostic genetic testing. A result from Nebula Genomics or Dante Labs — however technically accurate the underlying sequencing — has not been ordered, interpreted, or validated in a clinical diagnostic context. It is a consumer health product.

This matters for several reasons. If a consumer test identifies a variant that concerns you — particularly in a gene associated with cancer, cardiac disease, or another serious condition —

clinical confirmation through a CLIA-certified laboratory (in the US) or an UKAS-accredited NHS laboratory (in the UK) is the appropriate next step before any medical decisions are made.

A consumer WGS result is best understood as an informed starting point for a clinical conversation — not as the clinical evidence itself. The two things work together, not in place of each other.

What We Covered in This Chapter

Here are the ideas to carry forward:

- Whole genome sequencing reads every one of your roughly three billion base pairs — unlike SNP array tests, which check less than 0.03% of the genome. The cost has fallen from three billion dollars in 2003 to under three hundred dollars today for consumer 30× sequencing.
- 30× coverage — meaning each position in the genome has been read thirty times independently — is the clinical gold standard for accuracy. Look for this specification when evaluating any WGS service.
- The interpretation gap is the key limitation. A sequence is only as useful as the science available to interpret it. The majority of variants in any whole genome sequence will be classified as variants of uncertain significance — not because they are dangerous, but because science has not yet studied them in enough people to establish their meaning.
- The major consumer providers — Nebula Genomics, Dante Labs, Sequencing.com, and Invitae for clinical-grade testing — each make different trade-offs between price, reporting depth, privacy, and clinical credibility. Data download rights should be a primary consideration when choosing.
- The 23andMe bankruptcy of 2025 demonstrated

the importance of downloading and independently storing your raw data. A company's fate and your data's fate should not be the same thing.

- A whole genome sequence is most clearly valuable for people with unexplained family history of genetic conditions, rare disease situations, or those thinking long-term about having the most complete genomic baseline possible.
- Download your VCF file, store it in at least two places, tell your doctor you have it, and plan to reanalyze it periodically as the science advances. Your sequence does not expire — the knowledge to read it will keep growing.

Coming up in Chapter Six — Privacy, Data, and Who Owns Your DNA. The ethics and fine print of genetic databases who can access your genomic data, what legal protections exist, and how to make informed decisions about sharing the most intimate information you will ever generate.

+ + +

Chapter Six

Privacy, Data, and Who Owns Your DNA

The Ethics And Fine Print You Should Know Before You Test

This chapter draws on the work of The Kinnara Fund, a non-profit dedicated to the ethics of biotechnology. The questions raised here are ones the Fund has examined closely — because the right to understand who holds your biology, and on what terms, is not a technical question. It is a human one.

In June 2025, the US Senate Judiciary Committee convened a hearing titled, with admirable directness: '23 and You: The Privacy and National Security Implications of the 23andMe Bankruptcy.' Sitting before the committee were bioethicists, legal scholars, and privacy advocates. The question they had gathered to answer was one that many people across the world — people who had once spat cheerfully into a small plastic tube and sent it away in the post — were now asking themselves for the first time.

What exactly had they handed over? And to whom might it now belong?

This chapter is about that question. Not just about 23andMe

— that is one company's story, and we will tell it honestly — but about the broader landscape of genetic data: who collects it, what they do with it, what the law currently protects, where the law runs out, and what you can do to protect yourself when the law does not go far enough.

It is also about something more fundamental than legal protection. It is about the ethics of a new kind of information — one that is unlike any other kind of personal data we have ever generated. One that is permanent. One that cannot be changed if it is misused. One that, uniquely, does not belong only to you — because your genome contains information about your children, your parents, your siblings, your cousins. Decisions about sharing your genetic data are, whether you know it or not, decisions that affect people who never consented to be part of the conversation.

This is not a chapter designed to frighten you into never testing. The knowledge that genomic testing makes possible — the health insights, the medical applications, the family discoveries — is real and valuable, and the chapters before this one have tried to make that clear. This chapter is designed to ensure you enter that knowledge with your eyes open and your agency intact.

Your DNA is the most intimate data that exists. It predates you, it will outlast you, and it speaks for people beyond you. The decision about who holds it is worth making deliberately.

What Makes Genetic Data Different from Every Other Personal Data

When we talk about personal data privacy, most people are thinking about information like their location history, their browsing habits, their financial records, their messages. This is sensitive information, and its misuse can cause serious harm. But there is something qualitatively different about genetic data that places it in a category of its own.

First: it is permanent. Unlike a password, an address, a credit card number, or even a medical diagnosis, your genome cannot be changed if it is compromised. A data breach that exposes your genetic sequence exposes it forever. You cannot reset it.

Second: it is retrospectively and prospectively informative. Your genome contains information about your past — your ancestry, your biological family, your inherited traits — and your future — your disease risks, your drug responses, your biological vulnerabilities. No other form of personal data combines these two dimensions.

Third: it is relational. Your genome is, by definition, shared with your biological relatives. Approximately half of your DNA is identical to each of your parents and each of your children. Your siblings share, on average, fifty percent. Your first cousins share about twelve and a half percent. This means that when you share your genetic data, you are sharing information about people who did not consent to be included — people who may not even know you have tested.

A landmark paper published in Science in 2018 estimated that a genetic database covering just two percent of a population would contain a third-cousin match for virtually any person in that population. By 2025, the 23andMe database alone contained over fifteen million profiles. At that scale, a substantial fraction of the population of European descent is identifiable — not because they are in the database, but because someone related to them is.

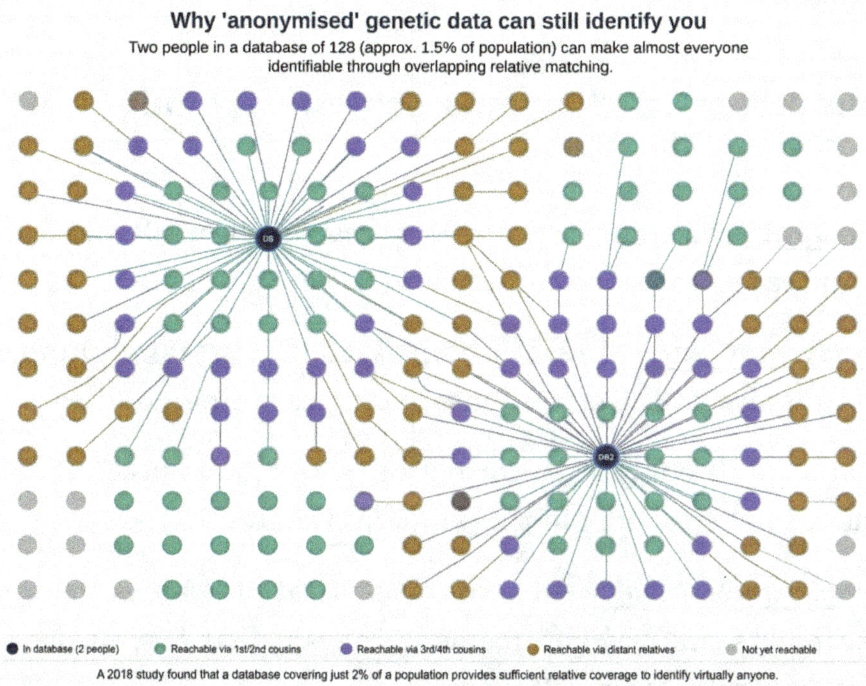

Figure 1: *The re-identification problem — two people in a database of figures, with connection lines showing how relative matching reaches most of the surrounding population. A database covering just 2% of a population contains enough relatives to identify the majority of that population's members.*
Source: Erlich et al., Science, 2018.

Fourth: it is commercially valuable in ways that are not always visible. Pharmaceutical companies, insurance companies, biotechnology researchers, and law enforcement agencies all have interests in genomic data. The business models of many consumer genomics companies depend on the monetization of their genetic databases — either directly, through data licensing agreements with research or commercial partners, or indirectly, through the research insights that genetic databases make possible. Your genome is not just private health information. In the commercial ecosystem of the twenty-first century, it is also an asset.

Four properties that make genetic data unique

Permanent — cannot be changed or revoked if misused. A compromised genome is compromised forever.

Doubly informative — reveals both past (ancestry, family) and future (disease risk, drug response, biological vulnerabilities).

Relational — contains information about biological relatives who did not consent. Sharing your data shares theirs.

Commercially valuable — to pharmaceutical companies, insurers, researchers, and law enforcement, in ways that may not be apparent at the point of collection.

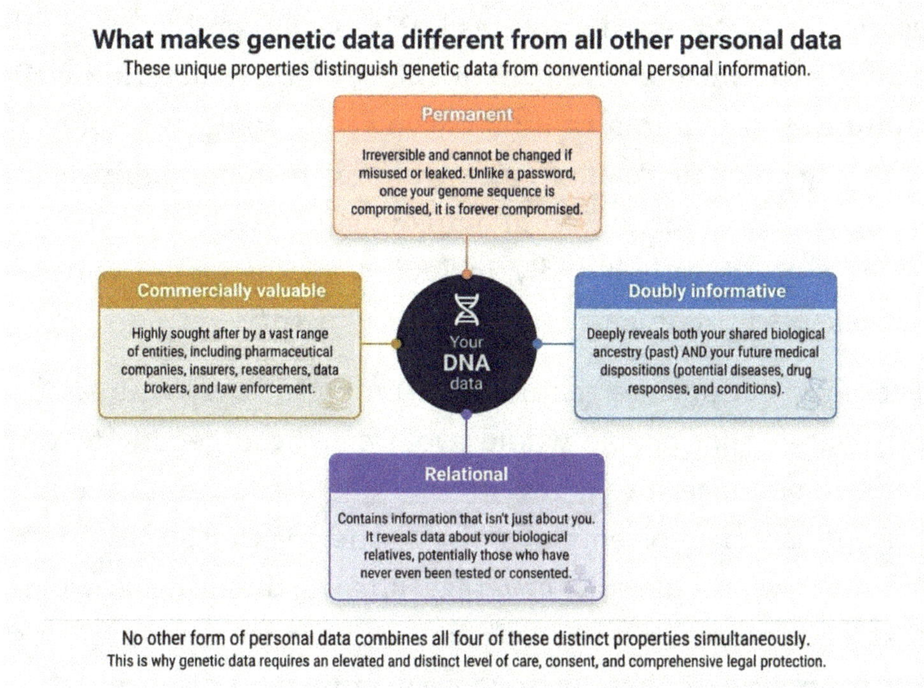

Figure 2: *Four properties that make genetic data unique — a radial diagram with four segments: Permanent (cannot be changed or revoked), Doubly informative (reveals past and future), Relational (implicates biological relatives), Commercially valuable (to pharma, insurers, law enforcement, and AI).*

The Legal Landscape: What Protects You, and What Doesn't

The legal protections for genetic data vary enormously depending on where you live, which type of entity holds your data, and what purpose it is being used for. Understanding these protections — and their gaps — is not optional if you are making an informed decision about genetic testing.

In the United States, the primary federal law governing genetic privacy is the Genetic Information Nondiscrimination Act, passed in 2008 and known as GINA. GINA prohibits health insurers from using your genetic information to determine eligibility, set premiums, or limit coverage. It prohibits employers with fifteen or more employees from requesting, requiring, or using genetic information in employment decisions. It defines genetic information broadly, including not just your own genetic test results but your family medical history.

These protections are genuinely meaningful. They addressed a real fear — that people would avoid beneficial genetic testing because of concern that results could be used against them in health insurance or employment — and for those specific domains, they provide solid coverage. It is worth noting, however, that legal protection and lived protection are not the same thing. For communities that already face documented discrimination in hiring, insurance underwriting, and healthcare access, the gap between what the law prohibits and what individuals can practically enforce is wider. GINA's protections matter most to those who have the resources and standing to invoke them.

But the gaps are significant, and they have been described by legal scholars as the law's most dangerous features.

Life insurance is not covered by GINA. If you apply for life insurance in the United States, the insurer can legally ask for your medical records — which may include genetic test results — and can use that information in underwriting decisions. Disability insurance and long-term care insurance are similarly unprotected. In early 2024, the life insurance company MassMutual announced a partnership with the genomics company Genomics plc, offering free genetic testing to its customers. The research implications for insurance underwriting are not difficult to infer.

HIPAA — the Health Insurance Portability and Accountability Act — protects the privacy of health information held by healthcare providers and health insurers. But consumer genomics companies are not healthcare providers. They are technology companies. HIPAA does not apply to 23andMe, AncestryDNA, Nebula Genomics, or Dante Labs. Their privacy practices are governed primarily by their own terms of service — terms that most people have never read, written in language that most people cannot easily parse, subject to change with notice that most people will never see.

As of 2025, twenty-two US states have enacted standalone genetic data privacy laws, with varying levels of strength. Alaska and Florida have laws that make genetic information the legal property of the person from whom it originated. Washington State's My Health My Data Act imposes regulatory restrictions on collectors of genetic data absent consent. California's Genetic Information Privacy Act provides some of the strongest consumer protections in the country. Colorado has enacted one of the most restrictive frameworks governing genetic data collection. But these protections are patchwork, inconsistent, and do not travel with you across state lines in any coherent way.

In the United Kingdom, the General Data Protection Regulation — incorporated into UK law after Brexit as the UK GDPR — treats genetic data as a special category of sensitive personal data, subject to the highest level of protection. Collecting and processing genetic data requires explicit consent and a clear legitimate purpose. The UK's Information Commissioner's Office has enforcement powers that the US federal government, with its fragmented regulatory framework, does not. For UK residents using consumer genomic services, the privacy protections are meaningfully stronger — but they apply to UK-based entities. A US company processing UK residents' genetic data may or may not be subject to full UK GDPR compliance, depending on where data is stored and processed.

The GINA gaps every consumer should know

Health insurance — PROTECTED. GINA prohibits health insurers from using genetic information to determine eligibility, set premiums, or limit coverage in employer-sponsored plans.

Employment — PROTECTED (employers with 15+ employees). Employers cannot request, require, or use genetic information in hiring, firing, or promotion decisions.

Life insurance — NOT PROTECTED federally. Insurers can ask for medical records and use genetic test results in underwriting. Several states have enacted additional protections — check your state law.

Disability insurance — NOT PROTECTED federally.

Long-term care insurance — NOT PROTECTED federally.

Military and federal government employees — GINA's employment protections do not apply to these groups.

Small businesses with fewer than 15 employees — GINA's employment protections do not apply.

Consumer genomics companies (23andMe, AncestryDNA, Nebula, etc.) — NOT covered by HIPAA. Governed by their own terms of service and applicable state laws only.

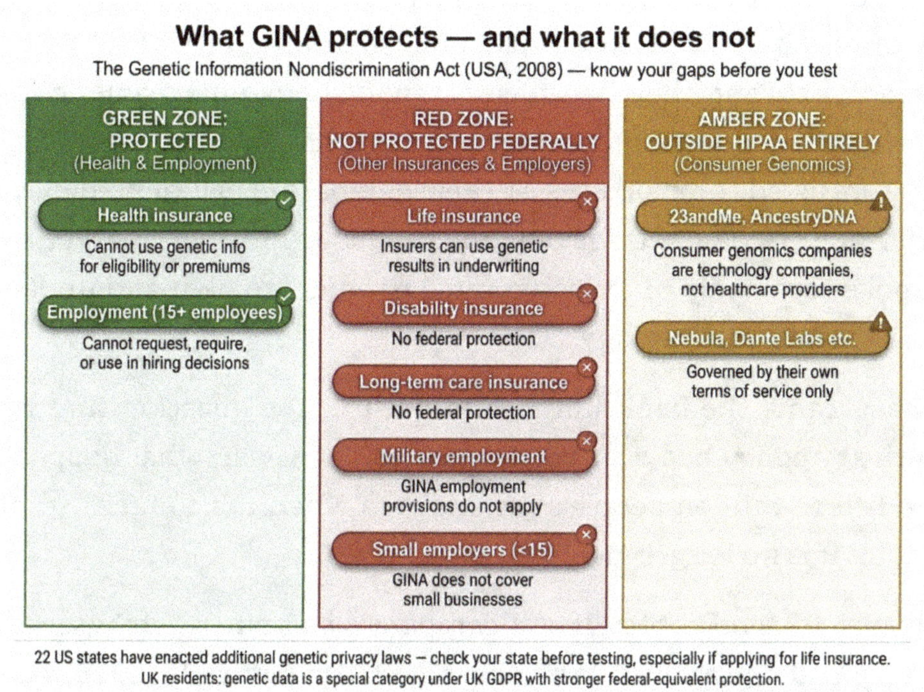

Figure 3: *What GINA protects and what it does not — three zones: Protected (health insurance and employment with 15+ employees), Not federally protected (life insurance, disability, long-term care), Outside HIPAA (consumer genomics companies governed by their own terms of service).*

The 23andMe Story: A Case Study in What Happens When Promises Meet Reality

In March 2025, 23andMe filed for Chapter 11 bankruptcy. What followed over the next three months illustrated, with unusual clarity, the gap between the assurances in a company's privacy policy and the legal reality of what those assurances are actually worth when financial pressure arrives.

23andMe had, over its nineteen-year history, built a reputation as one of the more privacy-conscious consumer genomics companies. Since 2015, they had published annual transparency reports stating they had never provided customer genetic information to law enforcement in response to a request. They had resisted forensic uses of their database. They offered clear opt-out mechanisms for research data sharing. Their marketing spoke repeatedly of empowering customers to own and understand their own biology.

Then came the bankruptcy, and with it, the question that no privacy policy had ever been designed to answer: what happens to fifteen million people's genetic data when the company that holds it is no longer viable as a business?

Under US bankruptcy law, a company's customer database can be considered a valuable business asset. On March 28, 2025, a US bankruptcy judge authorized 23andMe to include its customer genetic and survey data in its asset sale process. The California Attorney General issued a consumer alert urging 23andMe users to delete their data before any acquisition closed. Scholars from Harvard Law School published an analysis in the New England Journal of Medicine arguing that existing law failed to protect genetic data against exploitation during bankruptcy proceedings. The Senate Judiciary Committee scheduled a hearing.

The eventual resolution was, from a certain angle, reassuring:

the winning bidder was TTAM Research Institute, a nonprofit created by 23andMe's own founder, Anne Wojcicki. The genetic database did not pass to a pharmaceutical company, a foreign government, or an insurance conglomerate. But legal scholars and bioethicists were quick to note that this outcome was not the product of legal protection — it was the product of circumstances. The next time a consumer genomics company goes bankrupt, the buyer may not be so familiar.

Public Citizen, the consumer advocacy organization, noted that the Bankruptcy Code's protections for personally identifiable information do not expressly include genetic data — a loophole that Congress had still not closed as this book went to press, despite the Don't Sell My DNA Act being introduced in Congress in mid-2025. The Act would amend the Bankruptcy Code to require explicit consumer consent before genetic data could be sold or transferred in bankruptcy proceedings. Its passage, as of the time of writing, remains uncertain.

The 23andMe story matters not because 23andMe was uniquely careless. It matters because it demonstrated that even a company with genuine privacy commitments cannot guarantee the permanence of those commitments in the face of corporate failure. The privacy policy you agreed to exists at the pleasure of the company that wrote it.

A privacy policy is not a legal right. It is a company's promise — valid only as long as the company chooses to keep it, and legally enforceable only within very specific and often quite narrow limits.

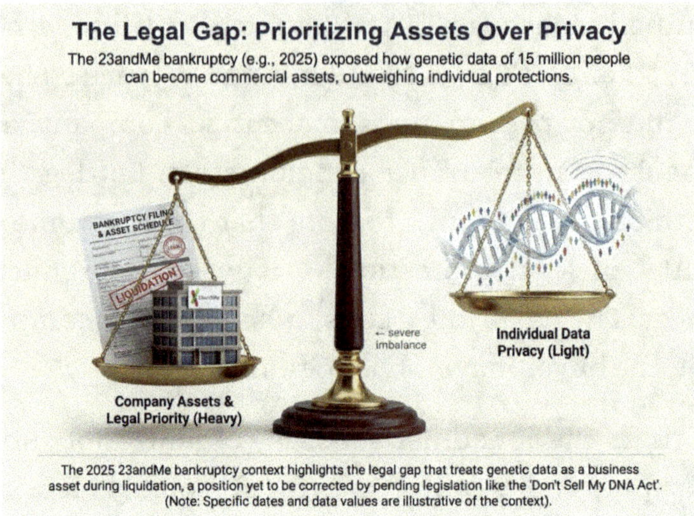

Figure 4: *A privacy policy is not a legal right — scales of justice with a DNA strand and a crowd of people on one side, legal and commercial interests on the other. The 23andMe bankruptcy of 2025 demonstrated that a privacy policy exists only as long as the company that wrote it continues to exist in its current form.*
 Commissioned for The Biology of You series.

The Research Consent You May Have Already Given

When you registered your consumer genomics test, you were almost certainly presented with one or more consent options relating to research. The framing varies by company, but the underlying structure is similar across the industry: would you like your anonymized genetic data to be used in research that could advance medical science?

This is not a cynical manipulation. The research value of large-scale genetic databases is genuine and significant. The NIH's All of Us Research Program, which has whole-genome sequences combined with medical records from approximately 400,000

US residents, is generating insights into disease biology that smaller studies cannot produce. 23andMe's own research partnerships, including a major collaboration with GlaxoSmithKline, have contributed to the identification of genetic variants relevant to drug development. The aggregate social benefit of people sharing their genomic data for research is real.

The ethical question is whether the consent people give to participate in this research is genuinely informed. Legal scholars have described consumer genomics terms of service as 'click-through contracts' — agreements that bundle sweeping data rights into documents that most people do not read, written in language most people cannot easily understand. The average consumer who clicks 'I agree' to research participation cannot reasonably anticipate every future use of their data: which pharmaceutical companies might license it, what research questions it might be applied to, what future analytical capabilities might reveal in a dataset that was generated years earlier.

Medical research ethics has long-established standards for informed consent. When a patient participates in a clinical trial, they receive detailed information about the purpose of the research, the potential risks, who will have access to their data, and their right to withdraw. The consent process is regulated, reviewed by an ethics board, and specifically designed to ensure that people are not consenting to something they do not understand.

Consumer genomics consent operates in a different world. The consent is typically given once, at the point of purchase, in a registration flow that most people complete in under three minutes. It is binding in ways that most people do not appreciate. And it often applies not just to specific studies, but to a broad class of future research purposes that are not defined at the point of consent.

Some companies are better than others. Nebula Genomics gives users granular control over data sharing, including the ability to opt into sharing with specific research partners individually and to revoke access at any time. Their blockchain-based consent architecture is one of the most transparent in the industry. But most consumer genomics companies operate closer to the other end of the spectrum: broad consent, limited visibility into how data is used, and a deletion process that may not remove data already used in completed research.

What 'anonymized' really means — and why it matters less than you think

When companies say your data is anonymized or de-identified before being shared with research partners, many people assume that means their genetic information cannot be traced back to them. This assumption is incorrect.

Genetic data is, by its nature, extremely difficult to truly anonymize. A 2018 study published in Science demonstrated that individuals could be identified from supposedly anonymous genetic datasets when a close relative had contributed their own data to a publicly accessible database. As genetic databases grow, the probability of re-identification from 'anonymous' data increases.

Under HIPAA, de-identified data is not covered by the regulation's privacy protections. This means that once your genetic data has been de-identified and shared with a research partner or third party, the original protections that governed your relationship with the testing company no longer apply — even if that data is later re-linked to your identity through methods that were not available when the data was first shared.

This is not hypothetical. The re-identification of 'anonymous' genomic datasets has been demonstrated in peer-reviewed research multiple times. It is a structural feature of the data type, not a recoverable edge case.

Law Enforcement, Forensic Genealogy, and the Database You Didn't Know You Joined

In April 2018, investigators in Sacramento, California, uploaded a DNA profile from a crime scene — evidence from a series of murders and sexual assaults committed between 1974 and 1986 — to GEDmatch, a publicly accessible genealogy database originally built by hobbyists to help adoptees and genealogists find biological relatives. The investigators were not looking for the perpetrator directly. They were looking for his relatives. Using the partial matches they found, they built a family tree. They narrowed it to a geographic region. They found a suspect. They collected a discarded coffee cup for confirmation.

The suspect was Joseph James DeAngelo — the Golden State Killer. He was arrested within days.

The technique used to identify him, known as investigative genetic genealogy or forensic genealogy, has since been used to identify suspects in hundreds of homicide and sexual assault cases and to identify hundreds of sets of unidentified human remains. It is a genuinely powerful tool for justice. It has also exonerated people wrongly convicted of crimes.

It has also never been the subject of comprehensive federal regulation in the United States.

The mechanism works because of the relational nature of genetic data. You do not need to be in the database to be identified. If a third or fourth cousin — someone you may never have met, whose name you might not know — has uploaded their DNA to a genealogy platform that allows law enforcement access, your biological identity can potentially be inferred from theirs. By 2025, it had been estimated that a database covering roughly two percent of a population provides sufficient relative coverage to identify virtually any individual in that population from a DNA sample.

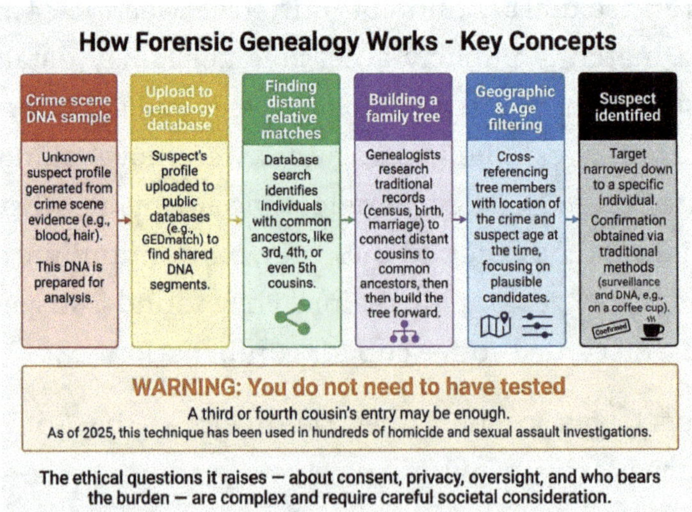

Figure 5: *How forensic genealogy works — six steps: crime scene DNA uploaded to a genealogy database, distant relative matches identified, family*

tree reconstructed, geographic and age filtering applied, suspect identified. The technique can identify individuals who never submitted their own DNA and never consented to participate in a forensic identification system.

GEDmatch, originally a volunteer-run genealogy site, was acquired in 2020 by Verogen, a forensic genomics company, which was itself later acquired by Qiagen in 2023. The acquisition brought the database's law enforcement access question from the realm of policy into the realm of corporate strategy. FamilyTreeDNA, another genealogy platform, acknowledged in 2019 that it had provided access to the FBI without disclosing this to its users — a revelation that prompted significant backlash from the genealogical community and a revision of the platform's terms.

23andMe, to its credit, maintained a strict policy prohibiting forensic uses of its database and published annual transparency reports confirming it had never provided customer data to law enforcement. But as the bankruptcy proceedings demonstrated, policies that exist at the company level exist only as long as the company exists in its current form.

The ethical dimensions here extend well beyond the question of whether forensic genealogy should be allowed. They include: who decides which databases can be used for law enforcement purposes? What consent should be required? What oversight should govern the practice? Who bears the burden of a legal framework that allows your biology to implicate your relatives without any of you having agreed to participate in a forensic identification system?

There is a dimension to the forensic genealogy question that deserves explicit acknowledgement. The CODIS database — the FBI's national forensic DNA index, which forms the backbone of law enforcement DNA identification in the United States — is not racially neutral in its composition. Because CODIS draws profiles predominantly from convicted offenders and arrestees, and because the US criminal justice system disproportionately arrests and convicts people of color, the database significantly overrepresents Black Americans and Latino individuals relative to the general population. The consequence of familial searching against a racially skewed database is that the families of people of color are reached by forensic genetic genealogy at a disproportionately higher rate. This is not a theoretical risk. It is a structural feature of how the system currently operates. Since 2020, a Department of Homeland Security rule has required DNA collection from immigration detainees, adding close to a million new profiles annually to CODIS — almost entirely people of color, the vast majority Latin — compounding the existing disparity. For readers from these communities, the forensic genealogy section of this chapter is not an abstract ethical discussion. It is a description of a system that is statistically more likely to implicate you or your relatives than it is to affect the average white American consumer of genetic testing services.

These are not abstract questions. They are questions about the design of a society that has embraced genetic testing before building the ethical and legal infrastructure to govern it responsibly.

The Kinnara Fund perspective: ethics does not wait for regulation

Law lags behind technology. This is not a failure of governance — it is a structural feature of how innovation moves faster than democratic deliberation. In genomics, the gap between what is technically possible and what is legally regulated has been particularly pronounced.

This does not mean that the absence of law implies the absence of ethical obligation. Companies that build genetic databases make choices — about consent architecture, data sharing agreements, law enforcement cooperation, security standards, and bankruptcy protections — that have real consequences for real people. Those choices are ethical choices, regardless of whether regulation compels them.

The Kinnara Fund's position is that individuals have a right to genetic privacy that precedes and exceeds what current law guarantees. That right includes the right to understand, in plain language, what will happen to their biological information — not just today, but in all the foreseeable futures that a company might navigate. And it includes the right to meaningful consent: not a click-through box, but an informed, specific, revocable agreement.

Advocacy matters. The Don't Sell My DNA Act. State-level genetic privacy laws. International data protection frameworks. These things exist because people and organizations pressed for them. If the legal protections you want do not yet exist, the path to them runs through engagement — with your elected representatives, with the companies whose services you use, and with the public conversation about what genomic ethics should look like.

The Commercial Ecosystem: Who Profits From Your Genome, and How

Understanding the business models behind consumer genomics is not a cynical exercise. It is a practical one. Companies that collect genetic data are not charities. They have revenue models. And those revenue models involve your genome in ways that are sometimes explicit and sometimes not.

The most transparent model is the direct sale: you pay for a test kit, you receive a report. In this model, the primary transaction is between you and the company, and your data's value to the company is in attracting you as a customer.

The second model is the research data partnership. In this model, the company's primary long-term revenue comes not from individual test kit sales — which, at $100 to $300 per kit, do not support the infrastructure required to maintain a database of millions of genomes — but from licensing anonymized genetic data to pharmaceutical companies, research institutions, and biotechnology firms for drug development and disease research. 23andMe signed a $300 million research partnership with GlaxoSmithKline in 2018. Nebula Genomics offers users the ability to earn compensation for sharing their data with specific research partners. Genomics England has data access arrangements with a range of pharmaceutical and research partners. These partnerships are, in many cases, clearly disclosed — but disclosed in terms of service documents that most consumers have not read.

The third model is emerging: genetic data as an input to AI and large language model development. As pharmaceutical companies and AI-driven drug discovery platforms invest in understanding the relationship between genetics and disease at scale, the value of large, diverse, well-annotated genetic databases continues to increase. The concern raised by bioethicists and legal scholars in 2025 is that consent frameworks designed in the 2010s — for genomic research as it existed then — may not adequately cover the kinds of analytical uses that 2025-era AI makes possible.

None of this means that the research use of genetic data is inherently wrong. The generation of pharmaceutical insights, the identification of disease mechanisms, the development of better-targeted treatments — these are goods that benefit people who will never have taken a genetic test. The question is not whether research should happen, but whether the people whose data makes it possible have given genuinely informed consent and whether the governance frameworks that oversee that research are adequate for the scale and sophistication of the work being done. The honest answer, in 2025, is that in many cases they have not and they are not.

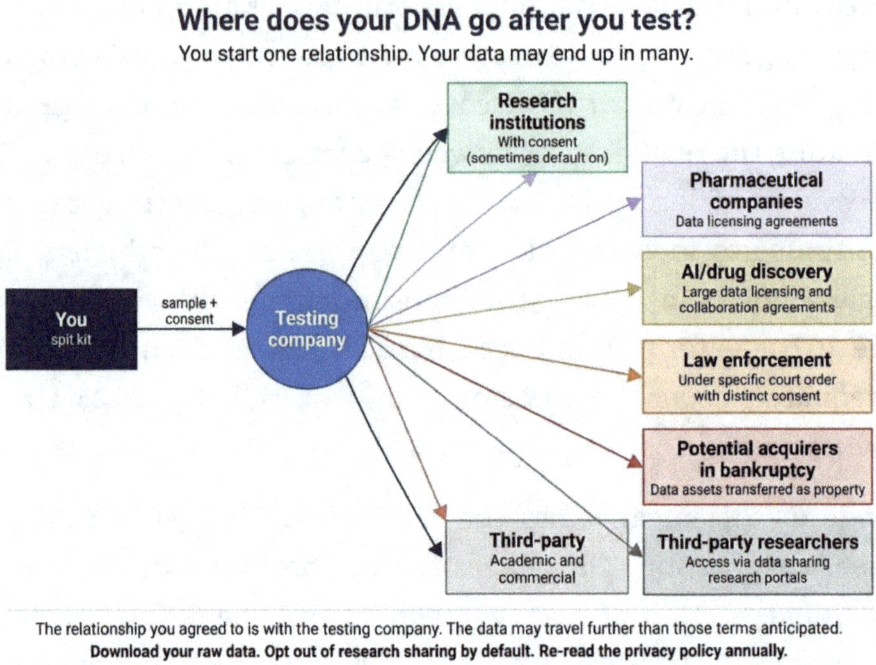

Figure 6: *Where your DNA goes after you test — a hub-and-spoke diagram from testing company to six potential destinations: pharmaceutical research partners, AI training datasets, law enforcement (via court order or voluntary policy), third-party researchers, potential future acquirers, and your own downloaded file. Only the last destination is fully within your control.*

The Equity Dimension: Who Bears the Risks?

There is a dimension to genetic data privacy that is rarely discussed in consumer genomics marketing materials, and that deserves space in any honest treatment of the subject.

The risks associated with genetic data are not distributed equally. Historically marginalized communities — Black Americans, Indigenous populations, immigrants, communities with histories of medical exploitation — have particular reason to approach genetic databases with care. The history of biological data collection from marginalized communities, often without meaningful consent and sometimes against active objection, is not distant or abstract. The Havasupai tribe's legal battle over the misuse of blood samples collected in the 1990s, ostensibly for diabetes research and used for purposes the community never consented to, is one of many examples.

The commercial genomics industry is not the same as the research institutions at the center of those historical abuses. But the structural conditions that made those abuses possible — power asymmetries, consent frameworks that favor the collector over the contributor, limited legal recourse for communities rather than individuals — persist in attenuated forms in the current consumer genomics landscape.

There is also the question of who benefits. The pharmaceutical and research insights generated from large genetic databases are, in their current form, predominantly generated from databases that over-represent people of European descent. Drug targets, disease risk models, and pharmacogenomic guidelines developed from these databases may work less well, or carry different risk profiles, for people from under-represented populations. The same communities that have the most reason for caution about contributing to genetic databases are also the communities whose exclusion from those databases produces the least representative, and potentially least equitable, science.

This is a genuine ethical dilemma, not a rhetorical one. Resolving it requires not better marketing but better governance: consent frameworks that are community-engaged, data-sharing models that return benefits to contributing communities, and regulatory oversight that treats the equity implications of genomic databases as a first-order concern rather than a footnote.

The question of who benefits from genetic databases, and who bears the risks, is not separate from the science. It is part of what the science is for.

Practical Steps: Protecting Yourself in the Current Landscape

The landscape described in this chapter is imperfect and will take years to improve through policy and advocacy. In the meantime, here is what you can do right now to navigate it with your interests protected.

Before you test, read the privacy policy of any service you are considering — specifically, look for answers to these questions. Does the company share data with third parties, and if so, who and under what conditions? Can you opt out of research data sharing without losing access to your health reports? What happens to your data if the company is acquired or goes bankrupt? What is the deletion process, and does deletion remove data from completed research? How long are physical samples retained, and can you request their destruction?

Opt out of research data sharing unless you have made a deliberate and informed choice to participate. This is not an anti-research stance — it is an act of meaningful consent. Research participation is more valuable when it is genuinely voluntary.

Download your raw data file immediately upon receiving results, and store it independently of the company's platform. Your data's value to you should not depend on the company's continued existence.

Consider using a pseudonymous email address for genomic testing services if you are concerned about data linkage between your genetic profile and other personal information held in databases linked to your primary email identity.

If you live in the United States, check your state's genetic privacy laws before testing, and understand whether your state's protections apply to life insurance underwriting. If you are currently applying for life insurance, disability insurance, or long-term care insurance, consider timing your genetic testing after those applications are complete.

If you have already tested and are uncertain about the status of your data, log into your account and review your current consent and sharing settings. Most major providers allow you to withdraw research consent and update data sharing preferences at any time — though withdrawing consent does not remove your data from completed studies.

If the data of your test has already been used in research you are uncomfortable with, you cannot undo that use. But you can delete your account, request destruction of your physical sample if it has not already been destroyed and document your request. Future protections may depend on having a clear record of having exercised whatever rights were available to you at the time.

The informed consumer checklist — before you test

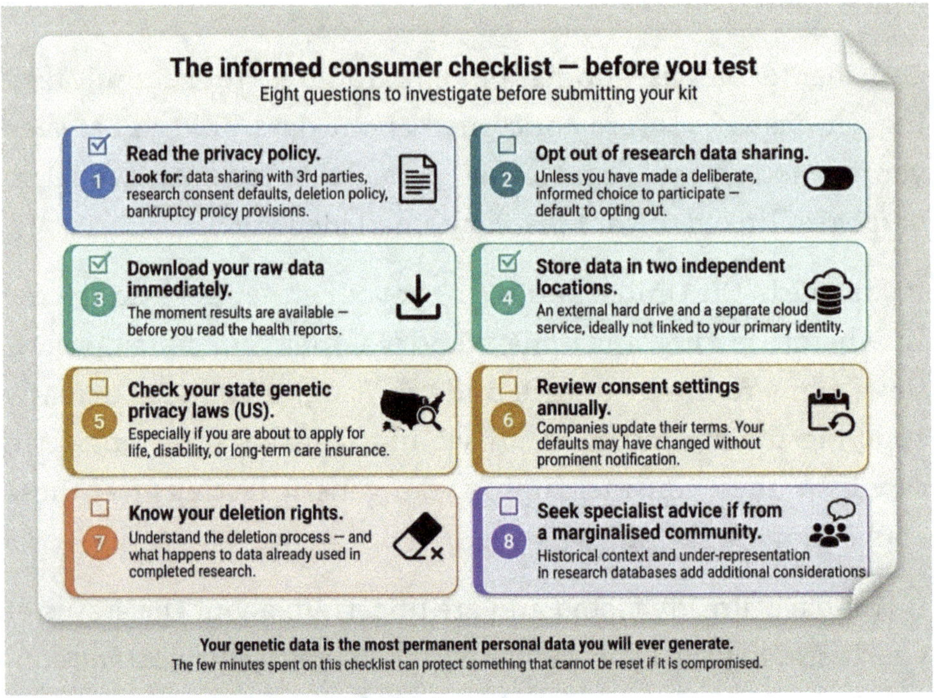

Figure 7: *The informed consumer checklist — eight actions before and after testing: read the privacy policy, opt out of research sharing, download raw data immediately, store in two independent locations, check state privacy laws, review settings annually, know your deletion rights, seek specialist resources if from a marginalized community.*

Read the privacy policy. Not all of it, necessarily — but look specifically for: data sharing with third parties, research consent defaults, bankruptcy and acquisition provisions, and deletion policy.

Opt out of research data sharing unless you choose otherwise deliberately.

Download your raw data file as soon as it is available.

Store your data separately from the testing company's platform, in at least two independent locations.

Check your state's genetic privacy laws if you are in the US. Understand the life, disability, and long-term care insurance implications.

Review your consent and sharing settings periodically — companies update their policies, and your settings may have been changed.

Know your deletion rights. Most providers allow deletion of your account and request for sample destruction — use them if you decide the tradeoff is no longer worth it.

If you are from a historically marginalized community, the following organizations offer specific resources, guidance, and advocacy on genetic privacy from a community justice perspective:

The Genetic Alliance (geneticalliance.org) provides plain-language guidance on genetic privacy rights and connects individuals with community-based genetic advocacy organizations.

The Global Indigenous Data Alliance (gida-global.org) and the CARE Principles for Indigenous Data Governance set out the rights of Indigenous communities over genetic and health data, and can help Indigenous individuals understand what protections and recourse exist for their data.

The ACLU's Racial Justice Program (aclu.org) has published extensively on genetic surveillance, CODIS racial disparities, and forensic genealogy's disproportionate impact on communities of color, and provides legal resources for individuals affected by these systems.

The Native BioData Consortium (nativebiodata.org) specifically addresses the intersection of Indigenous sovereignty and genomic research and provides guidance for Indigenous individuals navigating genetic testing decisions.

Before testing, consider whether the company you are considering has a stated policy on data sovereignty for Indigenous and marginalized communities, and whether it has made any public commitments to community benefit-sharing from research use of its database.

The Larger Conversation

Individual privacy practices matter. They are what you can act on today, with the landscape as it exists. But they are not sufficient, and they should not be mistaken for the whole answer.

The governance of genetic data is one of the defining ethical challenges of early twenty-first century technology. It involves questions about what consent means when the technology and its uses are evolving faster than people can meaningfully anticipate. It involves questions about property — whether your genome is a thing you own, or information about you, or something else entirely that existing legal categories do not capture well. It involves questions about the relationship between individual privacy and collective benefit, between the right to protect your own biological information and the potential social good of sharing it.

These questions do not have easy answers. But they have better and worse answers, and the difference between those outcomes depends substantially on whether the people affected — which is to say, everyone who has ever or will ever take a genetic test — are engaged in the conversation about what the rules should be.

The Don't Sell My DNA Act needs advocates. State-level genetic privacy laws need constituents to demand them where they do not exist and to strengthen them where they are weak. Consumer genomics companies need customers who read their terms of service and ask questions when those terms are inadequate. The researchers and bioethicists working on consent frameworks, equity in genomic data, and the governance of forensic genealogy need public support and public attention.

The biology of your genome was settled at the moment of your conception. What happens to the information that biology encodes is still, in many ways, unwritten. That is not a reason for resignation. It is a reason for engagement.

The most powerful thing an informed person can do with genetic knowledge is not just apply it to their own health. It is to participate in the conversation about how this knowledge is governed — for everyone.

What We Covered in This Chapter

Here are the ideas to carry forward:

- Genetic data is qualitatively different from other personal data: it is permanent, doubly informative (past and future), relational (it speaks for your biological relatives), and commercially valuable in ways not always disclosed at the point of collection.
- GINA protects against genetic discrimination in health insurance and employment, but leaves life insurance, disability insurance, long-term care insurance, and military employment uncovered. Consumer genomics companies are not covered by HIPAA.
- The 23andMe bankruptcy of 2025 demonstrated that privacy policies exist at the pleasure of the company that writes them. The Bankruptcy Code does not expressly protect genetic data — a gap the Don't Sell My DNA Act seeks to close.
- Research consent in consumer genomics is often broad, given at a single point, and not specifically informed about future uses. The 'anonymization' of genetic data provides far less protection than most people assume, because genetic data is structurally re-identifiable from relative matches.
- Forensic genetic genealogy — using consumer databases to identify criminal suspects through biological relatives — has solved hundreds of cold cases and operated largely without comprehensive federal regulation. You do not need to be in a database to be identified through it.
- The risks of genetic data exposure are not distributed

equally. Historically marginalized communities face particular risks and have particular reasons for caution. The science is also less complete for under-represented populations — participation and protection need to advance together.

- Practical steps include: reading privacy policies before testing, opting out of research sharing by default, downloading and independently storing your raw data, understanding your state's laws, and reviewing your consent settings periodically.
- Individual privacy protection matters but is not sufficient. The governance of genetic data is a collective challenge that requires policy engagement, advocacy, and public participation.

Coming up in Chapter Seven — Your Action Plan: What to Do With Your Results. A step-by-step guide from receiving your first test to building a personal genomic health practice working with your doctor, interpreting your data over time, and using your biological blueprint to make genuinely better decisions for your health.

✦ ✦ ✦

Chapter Seven

Your Action Plan

What To Do With Your Results — Now, And Over A Lifetime

You have spent six chapters learning to read the instruction manual inside you. You have come to understand what DNA is, how consumer tests work, what disease risk findings mean, how your genes shape your response to medications, what whole genome sequencing offers, and what the ethics of sharing your biological data require you to think about.

Now comes the part that matters most: what you actually do.

Knowledge without action is just anxiety with better vocabulary. The entire promise of genomic self-understanding — the reason this technology is worth knowing about, the reason the science that powers it is worth celebrating — is that it gives you the ability to make better decisions. About your health. About your screening. About your medications. About the conversations you have with your doctors. About the lifestyle choices that, as we established in Chapter One, have a genuine and measurable influence on whether genetic tendencies become biological realities.

This chapter is your practical guide. It is structured in three phases — immediate steps you can take as soon as your results arrive, a medium-term approach to building genomics into your regular healthcare, and a long-term framework for treating your genome as the living, evolving resource it is. And it gives significant space to the most valuable professional relationship you may not yet know you have access to: the genetic counselor.

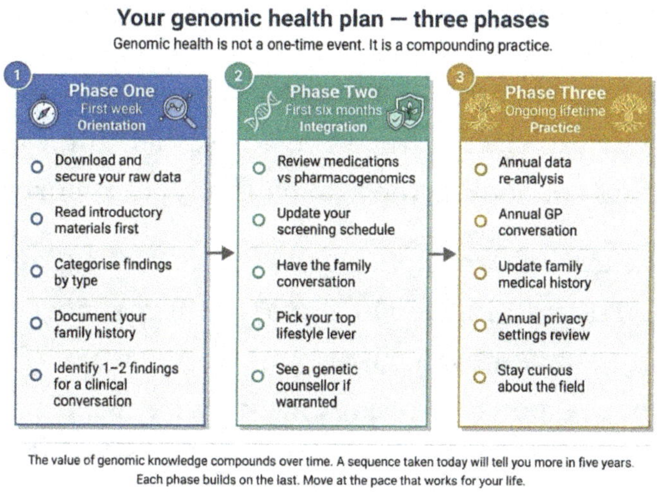

Figure 1: *Your genomic health plan — three phases across a timeline: Phase One (first week, Orientation), Phase Two (first six months, Integration), Phase Three (ongoing lifetime, Practice). Five numbered action steps per phase.*

Not everything in this chapter will apply to you. Your results are yours. Your family history is yours. Your healthcare system, your priorities, your circumstances — all of these shape what the right next steps look like. What this chapter gives you is a framework for deciding, rather than a prescription for everyone.

> *The most powerful use of genetic knowledge is not reading the report once and closing the browser. It is building a practice — a set of habits, conversations, and annual check-ins — that allows your biology to inform your life on an ongoing basis.*

Before Anything Else: The Right Mindset for Opening Your Results

If you have not yet opened your genetic results — or if you are planning a test in the future — there is something worth establishing before you look at the first page of your report.

Your results are information. Not a verdict. Not a diagnosis. Not a sentence. The framing you bring to the moment of reading will significantly shape how you respond to what you find — and whether that response is productive or whether it tips into the kind of anxiety that makes useful information feel like a burden.

The chapters before this one have tried to give you the context that transforms raw data into meaningful knowledge. A risk score means something specific: a statistical tendency in a population, modifiable by factors within your influence. A variant of uncertain significance means something specific: a genetic change whose effects are not yet established, not a hidden threat. A carrier status finding means something specific: relevant primarily for family planning, not for your own health.

With that context in hand, the mindset for opening results is one of curiosity rather than dread. You are reading a document about tendencies and possibilities. You are gathering information that, in many cases, will simply confirm what your family history already suggested. In some cases, you will find something that changes what you do — prompts a conversation with your doctor, triggers a screening test, shapes a medication decision. In very few cases will you find something genuinely alarming, and even then, as we established in Chapter Three, alarming information about genetic risk is, in virtually every case, information that opens options rather than closing them.

One practical suggestion: consider reading your results for the first time with another person present — a partner, a trusted friend, a family member. Not because you cannot handle the information alone, but because having someone with you changes the emotional register of the experience. It is harder to spiral into catastrophic interpretation when there is another person in the room asking sensible questions.

A checklist for before you open your results

Revisit the relevant chapters of this book for the result categories you are most expecting. If you know your family has a history of heart disease, re-read the BRCA and cardiovascular sections of Chapter Three. If you have been on antidepressants, re-read the pharmacogenomics section of Chapter Four.

Set aside an uninterrupted hour. Do not open your results on your phone during a commute or between meetings. Give the experience the attention and calm it deserves.

Have someone with you if possible, or have a trusted person you can call immediately after.

Download your raw data file before you read the report. Secure the asset before you engage with the content.

Know in advance who you will call if you find something that concerns you. Your GP's number. A genetic counselor's contact, if you have one. This is not pessimism — it is preparation.

Remind yourself, one more time: you are reading tendencies. You are not receiving a diagnosis.

Phase One — Immediate Steps

What to do in the first week after receiving your results

The first week after receiving your results is about orientation — understanding what you have, identifying the findings that need action, and beginning the conversations that will shape your next steps.

1. Download and store your raw data

Before you read a single health report, download your complete raw data file — at minimum your VCF, and your BAM and FASTQ files if your provider offers them. Store them on an external hard drive and in a separate cloud backup. Do this first. The reports can wait; your data should be secured immediately. Label the folder with the date and the testing company and record your account login details somewhere safe.

2. Read the introductory materials, not the risk scores

Most genomic testing platforms begin with educational content that explains what the tests measure and how to interpret results. Read this before diving into health sections. Fifteen minutes of context will save hours of unnecessary worry. Understanding that a polygenic risk score is a probability based on population statistics — not a prediction about you personally — fundamentally changes how a '1.4× elevated risk' figure reads.

3. Categorize your findings by type

Work through your report and sort findings into three categories. First, findings that are clinically established and potentially actionable — BRCA variants, familial hypercholesterolemia markers, DPYD status, APOE e4, high cardiovascular polygenic scores. Second, findings that are informational but not immediately actionable — trait reports, ancestry, low-moderate polygenic risk scores without family history corroboration. Third, variants of uncertain significance — note these, bookmark them, and file them for periodic re-review. Do not try to act on all three categories simultaneously.

4. Document your family medical history alongside your results

Your genetic findings and your family history are most meaningful when read together. A high cardiovascular polygenic score means something different if three of your four grandparents died of heart attacks in their fifties than if your family history is clean. Write down what you know of your parents', grandparents', and siblings' significant health conditions. This document will become one of the most valuable things you bring to a doctor's appointment.

5. Identify the one or two findings that most warrant a clinical conversation

Resist the temptation to take your entire report to your GP and ask them to explain all of it. Choose the one or two findings that sit in Category One — established, potentially actionable — and prepare a single focused question for each: 'My report shows I carry a BRCA2 variant. I understand this is a known risk variant for breast and ovarian cancer. I'd like to discuss what this means for my screening.' Specificity makes these appointments productive.

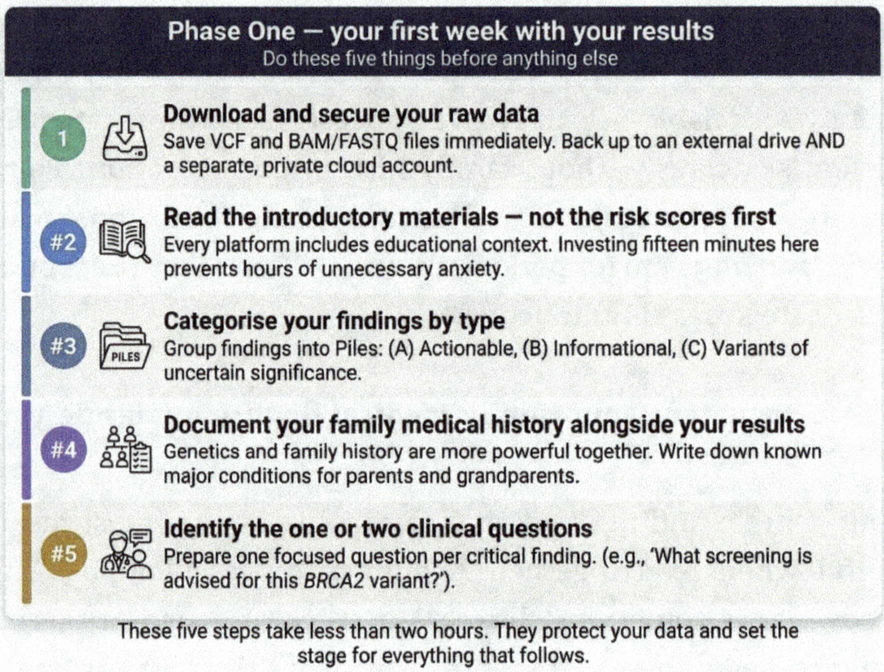

Figure 2: *Phase One checklist — five immediate steps: download raw data, read introductory materials before risk scores, categorize findings by type, document family history alongside results, identify one or two findings for a clinical conversation.*

* * *

Supporting guidance — working with a genetic counselor

Of all the professional relationships that genomic testing can open for you, the most valuable — and the least known — is the genetic counselor.

A genetic counselor is a healthcare professional with specialist training in medical genetics and counselling. Their role is not to tell you what to do with your genetic information. Their role is to help you understand what it means in the context of your personal and family history, to translate the language of variants and probabilities into the context of your actual life, to explain your options without imposing a particular course of action, and to connect you with the clinical resources you need.

Genetic counselors are trained to hold complexity. They understand that a BRCA2 result lands differently for a 28-year-old who has not yet had children than for a 55-year-old who has finished her family. They understand that APOE e4 means something different to someone whose mother is currently living with Alzheimer's than to someone with no personal experience of the disease. They can walk you through the decision tree for prophylactic interventions in a way that respects your values rather than simply reciting clinical guidelines.

They are also, crucially, trained to discuss the implications of your results for your biological family — who else might be affected by what you have found, how to think about sharing that information, and how to navigate the complex emotional terrain of genetic findings that affect people beyond yourself.

In the United Kingdom, genetic counselling is available through NHS clinical genetics services for people referred by a GP or specialist. In the United States, genetic counselors work in hospital genetics departments, cancer centers, prenatal clinics, and increasingly in private practice and via telehealth platforms. The National Society of Genetic Counselors maintains a searchable database of certified counselors at nsgc.org. In Australia, the Human Genetics Society of Australasia provides similar resources.

Consumer genomic testing companies do not, in most cases, provide genetic counselling as part of their service. 23andMe offers phone consultations with genetic counselors for customers who receive certain high-impact results — BRCA variants and the APOE e4 finding — but these are brief orientations, not the comprehensive sessions that a trained counselor in a clinical setting provides. If you receive a finding that concerns you, the phone consultation your testing company offers is a starting point, not a substitute for clinical genetic counselling.

Genetic counseling access — what to know if cost or geography is a barrier

Genetic counseling is more accessible than it used to be, but access is still uneven. Here is what is available if the standard pathways are not straightforward for your situation.

Insurance coverage: many US health insurance plans now cover genetic counseling when ordered by a physician for a clinical indication. If you have a significant genetic finding, ask your GP to make the referral — this typically brings it within insurance scope. Check your plan's benefits under 'genetic counseling' or 'medical genetics.'

Telehealth genetic counseling: platforms including Genome Medical (genomemedical.com) and Invitae's counseling service offer remote consultations that remove geographic barriers. Sessions typically run $200-$300 without insurance, and many are covered.

Equity-focused providers: Color Genomics (color.com) offers clinical genetic testing and counseling through employer health plans and community health programs, often at low or no cost. Their equity-focused model specifically aims to reach communities that have historically had limited access to genetics services.

Community health centers: Federally Qualified Health Centers (findahealthcenter.hrsa.gov) provide primary care on a sliding-scale basis and increasingly have genetics referral pathways. If you lack insurance, this is your most accessible entry point into the healthcare system for any follow-up.

NSGC financial assistance: the National Society of Genetic Counselors maintains information on financial assistance programs at nsgc.org. Some genetic counselors offer sliding-scale fees for uninsured patients.

When to seek a genetic counselor

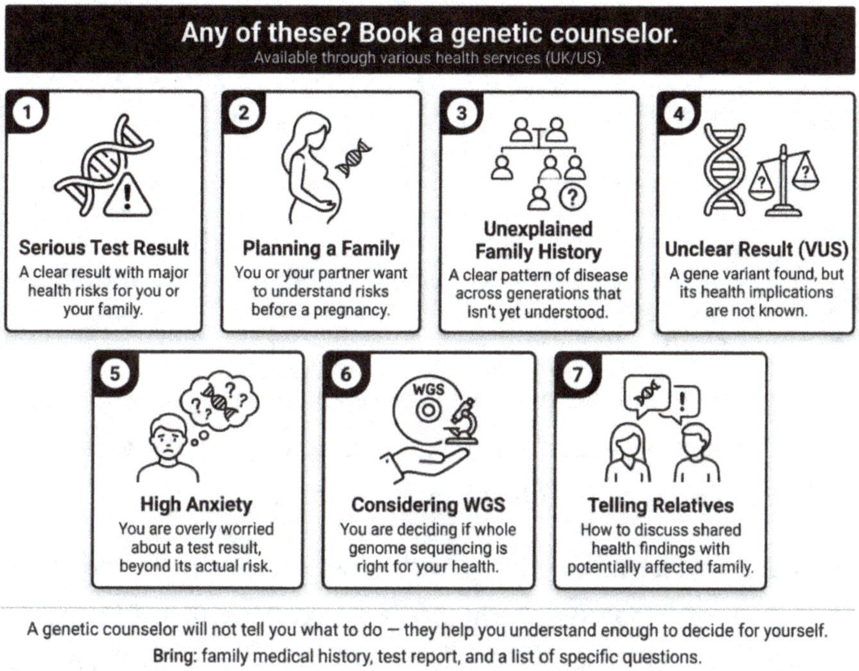

Figure 3: *When to seek a genetic counselor — seven situations listed as tiles: high-penetrance finding, family planning or carrier status, strong unexplained family history, VUS in a serious gene, disproportionate anxiety, WGS decision, sharing findings with family members.*

1. You have received a finding in a high-penetrance category — BRCA1/2, familial hypercholesterolemia, or another single-gene condition with well-established clinical implications.

2. You are considering having children and want to understand your carrier status for recessive conditions, or you have received a carrier status finding and want to understand its implications for family planning.

3. You have a strong family history of a genetic condition and want to understand your personal risk and what testing might be appropriate.

4. You have received a variant of uncertain significance in a gene associated with a serious condition, and you want help understanding whether it warrants monitoring or can be set aside.

5. You are finding that your genetic results are causing significant anxiety that is disproportionate to the clinical significance of the findings — a genetic counselor can help recalibrate the emotional response to the actual evidence.

6. You are trying to decide whether whole genome sequencing is appropriate for your situation, and you want guidance from someone who understands both the technology and your specific medical context.

7. You want to think through how to share genetic findings with family members who may be affected.

The first appointment with a genetic counselor typically lasts 45 to 90 minutes. It is helpful to bring your family medical history document, your genetic test report, any previous clinical genetic testing results, and a list of the specific questions you want answered. Come prepared to talk about your family — who has been affected by what conditions, at what age, and how. That context is what allows a genetic counselor to do their best work.

One thing worth knowing: genetic counselling is not therapy, and it is not a decision-making service. A genetic counselor will not tell you whether to have preventive surgery, whether to share your results with your family, or whether to take a particular medication. What they will do is make sure you understand your situation clearly enough to make those decisions yourself, with full information and a realistic sense of what the evidence supports. A full directory of genetic counseling resources for English-speaking countries is in Appendix C at the back of this book.

A genetic counselor is not a gatekeeper between you and your results. They are a guide between your results and your life — someone trained to help you translate molecular information into human decisions.

Supporting guidance — taking your results to your doctor

Many people who receive genomic test results are uncertain how to raise them with their GP or family doctor. Will the doctor take them seriously? Will they know how to interpret them? Will they dismiss a consumer test as unreliable and refuse to engage?

These are legitimate concerns. Primary care doctors vary enormously in their familiarity with genomics, and the honest reality is that many have received limited training in interpreting consumer genomic results. But the answer to this uncertainty is not to stay silent about your results. It is to come to the conversation prepared.

The most productive approach is to be specific, be concise, and be prepared to explain the context of what you found. Rather than arriving with a printed copy of your entire report and asking your doctor to evaluate it, identify the single finding you most want to discuss, explain what the test showed and what you understand it to mean, and ask one focused question: what would you recommend as a next step given this result?

A GP who is unfamiliar with pharmacogenomics can still order a formal CYP2D6 clinical test if you explain that your consumer report suggested you may be a poor metabolizer and you are about to start an antidepressant. A GP who has never seen a polygenic cardiovascular risk score can still order a lipid panel, check your blood pressure, and discuss your modifiable risk factors. They do not need to be experts in genomics to act on the clinical question your results have raised.

For findings that clearly require specialist input — BRCA variants, hereditary cancer syndromes, rare conditions identified through whole genome sequencing — your GP is a referral gateway rather than the end point. Ask explicitly for a referral to a clinical genetics service, a genetic counselor, or the relevant specialist. In the US, a referral to a genetic counselor or a medical geneticist is the appropriate path. In the UK, this is available through the NHS. Be persistent if you meet initial resistance — the evidence base for acting on these findings is strong enough to warrant specialist input.

How to prepare for a GP appointment about your genomic results

Write a one-paragraph summary of your most significant finding before you go. Describe what the test found, what category it falls into (high-penetrance variant, elevated polygenic score, pharmacogenomic finding), and what question you want answered.

Bring your family medical history document. The convergence of your genetic findings with your family history is what makes the clinical picture most legible.

Know the name of the gene or variant involved. 'My report flagged a BRCA2 variant' is a more productive opening than 'my test said I have an elevated cancer risk.'

Have a specific ask ready. 'I would like a referral to a clinical genetics service' or 'I would like a lipid panel and blood pressure review given my cardiovascular risk score' gives the doctor something concrete to respond to.

If your doctor is unfamiliar with consumer genomic testing, explain that the underlying genotyping technology is the same as that used in clinical SNP array testing. The consumer report is an interpretation layer on top of a scientifically validated test.

If you receive a dismissive response, you are entitled to a second opinion. Genomic medicine is advancing rapidly and the evidence base for acting on high-penetrance findings is unambiguous.

Phase Two — Building Genomics Into Your Healthcare

What to establish in the first six months

Once you have processed your immediate findings and begun the conversations they warrant, the next phase is about integration — making genomic information a standing part of how you think about and manage your health.

6. Review your medications against your pharmacogenomic findings

Work through every medication you currently take — prescription and over-the-counter — and check whether your pharmacogenomic results include any relevant findings. Pay particular attention to antidepressants, pain medications, anticoagulants, antiplatelet drugs, and any medication you have had trouble with in the past. If you identify a potential pharmacogenomic interaction, bring it to your prescribing doctor. This is one of the most immediately actionable areas of genomic medicine and one of the most frequently overlooked.

7. Establish or update your screening schedule

Use your genetic results to calibrate your preventive screening timeline. A high cardiovascular polygenic score may warrant earlier lipid testing and blood pressure monitoring than standard guidelines recommend. A BRCA variant warrants discussion of enhanced breast or prostate screening protocols. Even a moderate elevated risk for colorectal cancer may be worth discussing with your doctor in terms of when to begin colonoscopy screening. The goal is not to over-medicalized a healthy body — it is to direct available preventive resources toward the areas where they will do the most good.

8. Have the family conversation

If your results include findings that have implications for your biological relatives — a BRCA variant, familial hypercholesterolemia, a carrier status finding relevant to family planning — you now face one of the more humanly complex aspects of genetic testing: whether and how to share this information. There is no legal obligation in most jurisdictions to tell your relatives about your genetic findings. But there is a compelling ethical argument that family members who may be affected by a heritable condition have a genuine interest in knowing about it. A genetic counselor can help you navigate the conversation — both how to have it and how to support relatives who receive news they were not expecting.

9. Address your top lifestyle lever

Revisit your results through the lens of the modifiable factors we discussed throughout this book. If your results flag elevated cardiovascular risk, the most powerful single lever is almost certainly physical activity — study after study shows that regular aerobic exercise reduces the realized incidence of heart disease in high-genetic-risk individuals more than almost any other non-pharmaceutical intervention. If your results flag elevated risk of Type 2 diabetes, the evidence most strongly supports weight management and reducing refined carbohydrates. If APOE e4 is in your profile, cardiovascular health and quality sleep are the interventions with the strongest evidence base for brain protection. Pick one lever. Start there. Sustainable behavior change is built one habit at a time, not through wholesale lifestyle overhaul.

10. Discuss epigenetic testing if it is relevant to you

If you are interested in understanding your biological age relative to your chronological age — and in tracking whether the lifestyle changes you make are having a measurable effect on how your cells are aging — epigenetic age testing is now accessible through consumer platforms including TruMe, Elysium Index, and Tally Health. These tests measure methylation patterns on your DNA and calculate a biological age estimate. They are not perfect instruments, and the science of epigenetic clocks is still developing. But for people who are making deliberate lifestyle changes and want a way to measure their biological impact, they offer something no other currently available test can: a direct window into the epigenetic layer that sits above your genome. We explore this in depth in Book Five of this series.

Lifestyle Adjustments That Genomics Points You Toward: The Evidence Hierarchy

This section is not a generic wellness guide. Every intervention mentioned here has been selected because it has a specific, evidence-based relationship to the genetic risk categories most commonly found in consumer genomic reports. The goal is not comprehensiveness — it is precision. These are the levers most likely to matter, for the risks most likely to be flagged.

For elevated cardiovascular polygenic risk, the evidence on lifestyle is not just encouraging — it is striking. People in the highest genetic risk quartile for heart disease who maintained a healthy lifestyle cut their realized ten-year risk nearly in half compared to those with the same genetic risk who did not. That is not a marginal improvement. It is the difference between a high-risk trajectory and one that approaches the population average — achieved without any clinical intervention, through choices that are available to almost everyone. Physical activity sits at the top of the evidence hierarchy: 150 minutes or more of moderate aerobic activity per week is the single most consistently supported intervention. Below that: not smoking, managing blood pressure, maintaining a healthy weight. Diet matters, but its independent effect size is smaller than exercise and smoking cessation in most studies of cardiovascular genetic risk. Blood pressure management is particularly critical and is often underweighted by people who focus primarily on cholesterol. For readers of African descent specifically, blood pressure is the primary cardiovascular signal — as established in Chapter Three, populations of African ancestry tend to carry cardiovascular risk primarily through hypertensive pathways rather than cholesterol pathways. The lifestyle levers that reduce blood pressure — aerobic exercise, stress reduction, sleep quality, and dietary sodium management — are therefore the most directly relevant actions for that group.

For elevated Type 2 diabetes risk, the Diabetes Prevention Program — a landmark clinical trial that enrolled people at high risk of Type 2 diabetes — found that modest weight loss of five to seven percent of body weight, achieved through dietary modification and moderate exercise, reduced the development of diabetes by 58 percent over three years. Metformin, by comparison, reduced it by 31 percent. For people with high genetic risk who are also carrying excess weight, weight management is consistently the single most impactful intervention available.

For APOE e4 carriers, the evidence increasingly points to a cluster of interconnected behaviors rather than a single silver bullet. Cardiovascular risk management — blood pressure, physical activity, avoiding smoking — is relevant because vascular health and brain health are deeply linked. Sleep quality is emerging as particularly important: during deep sleep, the brain's glymphatic system clears amyloid beta — the protein that accumulates in Alzheimer's disease — from brain tissue. Seven to nine hours of quality sleep per night is not a luxury for an APOE e4 carrier; it is one of the most biologically direct things they can do. Cognitive engagement — learning new skills, maintaining social connection, intellectually challenging work — is associated in longitudinal studies with later onset of cognitive decline and larger 'cognitive reserve.'

For BRCA carriers who are managing risk through enhanced surveillance rather than preventive surgery, the evidence supports several risk-modifying lifestyle factors. Maintaining a healthy weight reduces breast cancer risk in general populations, and the relationship appears to hold in BRCA carriers. Limiting alcohol consumption is supported by strong evidence — alcohol is a direct carcinogen for breast tissue, and even moderate consumption raises risk. Physical activity has a modest but consistent risk-reducing association. These factors do not eliminate BRCA-associated risk, but they operate on top of the clinical surveillance and medical management strategies that are the primary tools for BRCA carriers.

The five lifestyle levers with the best evidence across multiple genomic risk categories

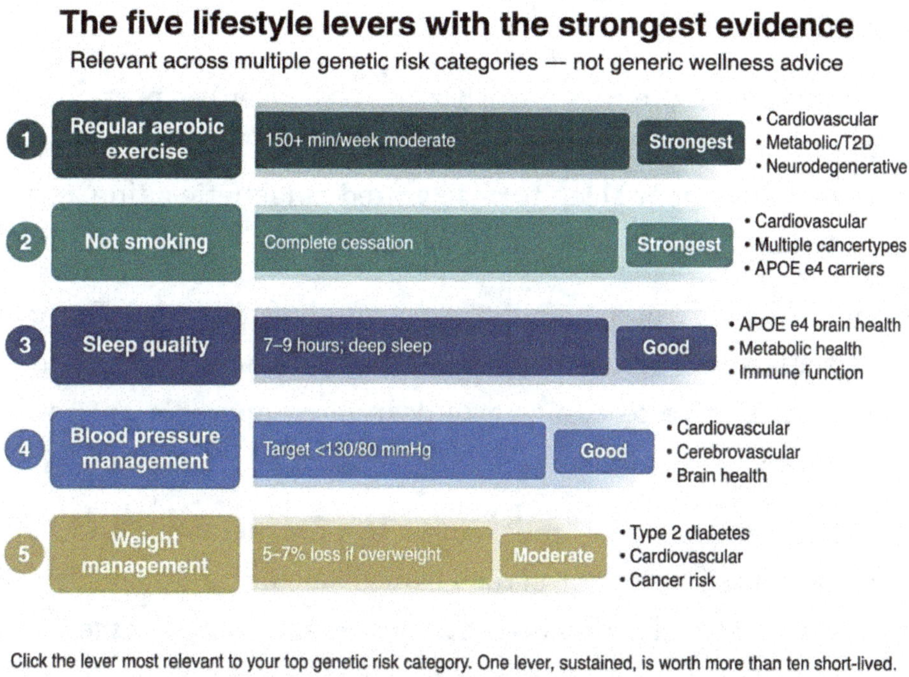

Figure 4: *The five lifestyle levers ranked by breadth of evidence — aerobic exercise, not smoking, sleep quality, blood pressure management, weight management. Each lever shows the genomic risk categories where its effect is best supported. Exercise and blood pressure management are highlighted as the two levers with the broadest documented impact across risk categories.*

Regular aerobic exercise (150+ min/week moderate, or 75+ min/week vigorous) — reduces realized risk in cardiovascular, metabolic, and neurodegenerative categories. Possibly the most broadly effective single modifiable factor across all genetic risk profiles.

Not smoking — reduces cardiovascular, pulmonary, and cancer risk across virtually all genetic backgrounds. For high cardiovascular polygenic scorers, smoking is the single most powerful risk amplifier.

Sleep quality (7–9 hours, prioritizing deep sleep) — particularly relevant for APOE e4 carriers and metabolic health. Emerging evidence links poor sleep to accelerated epigenetic aging across all risk categories.

Blood pressure management — strongly supported for cardiovascular and cerebrovascular risk reduction. Often underweighted relative to cholesterol despite comparable or greater effect sizes.

Weight management — most impactful for metabolic and Type 2 diabetes risk, relevant across cardiovascular and cancer risk categories. The relationship between weight and expressed genetic risk is consistent across a wide range of conditions.

Phase Three — A Lifetime Genomic Health Practice

Annual habits that make your genome work for you over the long term

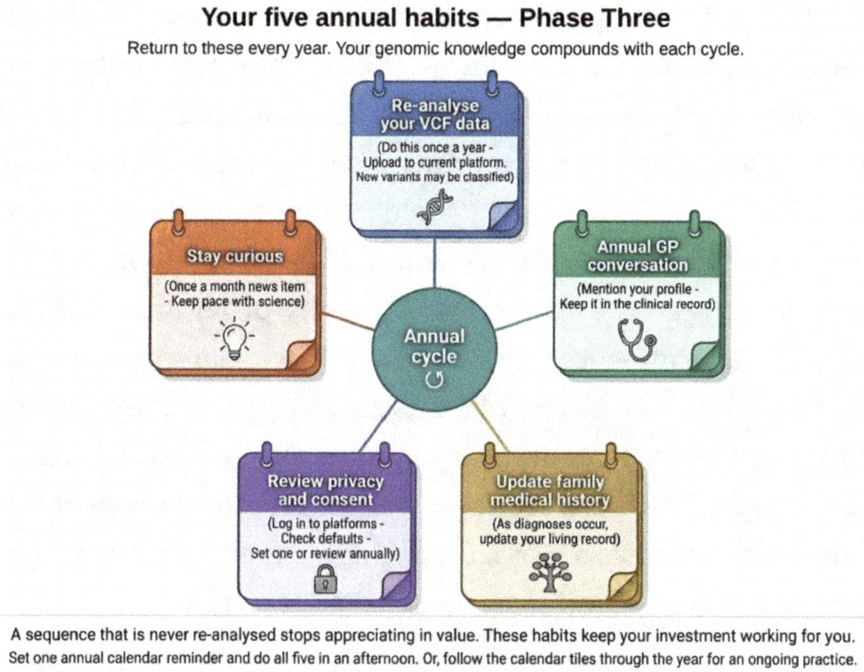

Figure 5: *Your five annual habits — a circular card for Phase Three: re-analyze VCF data with current tools, annual GP conversation mentioning genomic profile, update family medical history with new diagnoses, review privacy and consent settings on all platforms, stay curious about field developments.*

The final phase of your action plan is not about a single set of tasks to complete — it is about establishing a set of annual habits that allow your genomic knowledge to compound in value over time, as the science advances and as your own health picture evolves.

11. Annual genomic data re-review

Once a year, return to your raw VCF file and re-analyze it using current interpretation tools. Promethease and similar platforms can be updated with your existing data at any time. If you use a consumer testing platform with automatic report updates — Nebula Genomics, Dante Labs — check what new reports have been published since you last logged in. Science in this field moves fast enough that an annual review will, over a five-to-ten year horizon, consistently yield new findings from data you have already collected. This is especially true for people of non-European ancestry: as the major diversity programs — NIH All of Us, the African Ancestry Genome Project, GenomeAsia 100K — expand their databases, variants currently classified as uncertain for non-European ancestry genomes will be reclassified at a faster rate than for European ancestry genomes. The annual re-analysis habit has a particularly high expected return if your heritage is from an underrepresented population.

12. Annual conversation with your GP about your genomic profile

Your genomic findings are not a one-time conversation. Make it a habit to briefly mention your genetic risk profile at your annual check-up — particularly in relation to any screening tests that are due, any medications you are starting, or any new family history that has emerged. Keeping your doctor updated means your genomic information stays in the context of your evolving health picture rather than sitting in isolation in an app you opened once.

13. Track your family history updates

Your family medical history is a living document. As relatives are diagnosed with conditions, update your record. As you learn more about the health histories of older family members, add what you discover. The convergence of your genetic findings with your expanding family history is what gives both sources of information their maximum clinical value. A high polygenic cardiovascular score with no family history has a different clinical weight than the same score in a family where multiple members have had early heart attacks.

14. Review your consent and privacy settings annually

Once a year, log into every genomic testing platform where you have an account and review your data sharing and research consent settings. Companies update their terms of service. Default settings change. New data-sharing arrangements are entered into. This takes ten minutes per platform and is one of the most directly protective things you can do to maintain meaningful control over your biological information.

15. Stay curious

Genomics is one of the fastest-moving fields in science. The treatments being developed for BRCA carriers, the clinical trials exploring APOE e4 interventions, the pharmacogenomic guidelines being embedded in clinical care, the whole-genome sequencing becoming routine in NHS oncology — all of these are advancing in real time. Following a reputable source of genomic medicine news — the National Institutes of Health, Genomics England, the journal Nature Genetics, or the STAT News genetics section — means that your investment in genomic self-knowledge compounds with the field rather than becoming dated. You have done the work to understand the foundation. Keep building on it.

Follow-Up Testing: When to Go Further

Consumer genomic testing is a starting point, not a clinical conclusion. There are several situations in which a consumer result should prompt more comprehensive testing through a clinical laboratory or specialist service.

A consumer test that flags a BRCA1 or BRCA2 variant should be confirmed by a clinical diagnostic test before any medical decisions are made. Consumer SNP array tests check for a specific list of known BRCA variants — they do not read the entire gene. A clinical test that sequences the complete gene may identify additional variants the consumer test missed, or may confirm that the consumer result is accurate. This confirmation is not bureaucratic formality — it is essential, because the clinical interventions available to BRCA carriers are significant, and they should be based on the most complete information available.

A consumer pharmacogenomics result that suggests you are a poor or ultra-rapid metabolizer for CYP2D6 — particularly if you are taking medications in that enzyme's portfolio — is best confirmed by a dedicated clinical pharmacogenomics panel. As we discussed in Chapter Four, CYP2D6 is a structurally complex gene that consumer SNP arrays do not fully characterize. A clinical panel uses more comprehensive methods and provides a more complete picture.

A consumer test that identifies variants of uncertain significance in genes associated with serious conditions is a reason to have those specific variants reviewed by a clinical geneticist or genetic counselor, who can assess the variant's location, population frequency, and available evidence in a way that a consumer report interface cannot.

If a consumer test does not identify a cause for a strong family history pattern — multiple relatives with the same cancer type, early-onset cardiac disease, unexplained neurological conditions — it does not rule one out. Consumer SNP arrays miss a substantial proportion of pathogenic variants simply because they are not located at the positions the array checks. Whole genome sequencing, or targeted clinical panel testing for the condition in question, is the appropriate next step when family history strongly suggests a genetic cause that consumer testing has not found.

A note on direct-to-consumer testing and clinical confirmation

Consumer genomic tests are not clinical diagnostic tests. They are health products built on solid technology, but designed for a consumer audience rather than a clinical one.

For people of non-European ancestry, there is an additional reason to seek clinical confirmation for any significant finding. Consumer testing panels were built predominantly from European population data. Clinical panels draw on broader and more diverse reference databases, and are more likely to give you an accurate, complete picture of what a variant means for someone with your specific ancestry. If a consumer test flags something that concerns you, clinical confirmation is not just good practice — for non-European ancestry individuals it may be the difference between a meaningful result and a misleading one.

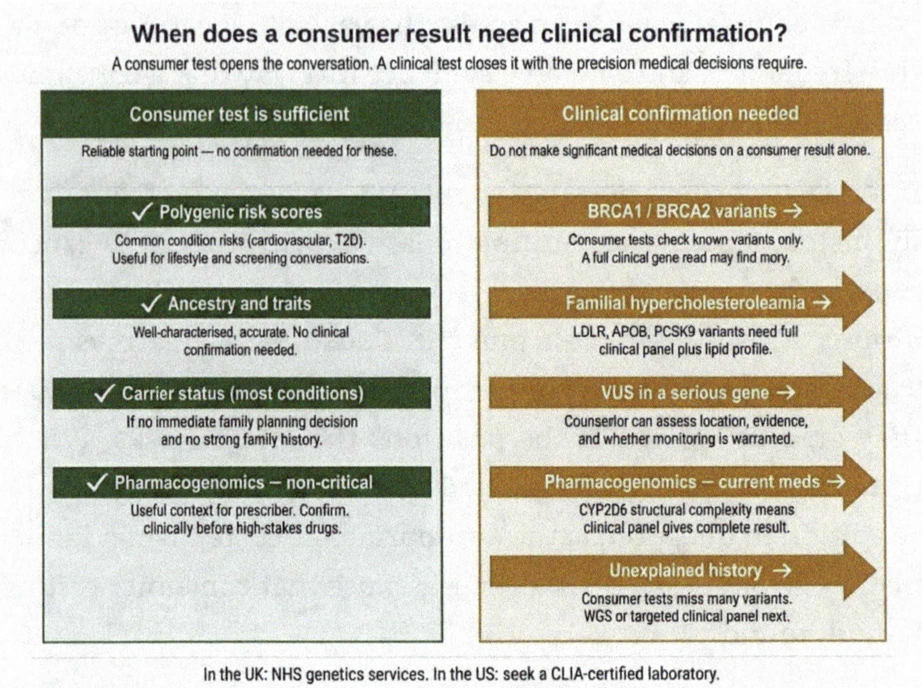

Figure 6: *Consumer test vs clinical confirmation — two columns: Consumer test sufficient (polygenic scores, ancestry, most carrier status, non-critical pharmacogenomics) vs Clinical confirmation needed (BRCA, FH, VUS, current medication conflicts, unexplained family history).*

This does not mean consumer test results are inaccurate. For common, well-characterized variants, they are generally reliable. It means they are incomplete — they check specific known positions and report on them, leaving the broader genetic landscape unexamined.

For any finding that might change a clinical decision — a treatment choice, a surgical consideration, a family planning discussion — clinical confirmation is the appropriate standard. A consumer test opens the conversation. A clinical test closes it with the degree of precision that medical decisions require.

The Conversation This Book Was Always Building Toward

In Chapter One, we began with a simple idea: you have an instruction manual inside you that has been shaping your life since before you were born, and you have never been taught to read it.

Seven chapters later, you can read it. Not every word — nobody can, yet. The science is still filling in what the three billion letters mean, one study at a time, one population at a time. But you can read the sections that matter. You can understand your risk variants without catastrophizing them. You can engage with pharmacogenomic findings and bring them to your prescriber with confidence. You can hold the APOE e4 result with equanimity rather than dread, because you know what it means and what it does not. You can navigate the privacy landscape with eyes open. You can take your genome to your doctor as a tool rather than a mystery.

What changes when people understand their biology is not just the decisions they make. It is the quality of the relationship they have with their own health — less passive, less anxious, more directed, more curious. It is the quality of the conversations they have with their doctors — more specific, more informed, more equal. It is the quality of the choices they make about lifestyle, screening, medications, and family planning — more grounded in evidence, less driven by fear or by hope without foundation.

None of this is simple. Genomics is a field that rewards careful thinking, honest engagement with uncertainty, and a genuine willingness to hold complexity. The instruction manual inside you is not a simple document. But it is yours. And understanding it — even partially, even imperfectly, even in the knowledge that much of it remains unreadable for now — is one of the most genuinely empowering things you can do for your health.

The series continues with books on the microbiome — the universe of microbes inside you that shapes your immunity, metabolism, and mental health — and on gene editing, the technology that is beginning to rewrite the code this book has taught you to read. Further books in The Biology of You series will cover longevity science, personalized nutrition, regenerative medicine, and the role of artificial intelligence in your health data.

Each book builds on the foundation this one has laid. The more of the series you read, the more the pieces connect. But start where you are. With what you know now. With the results you have, or the test you are about to take, or the conversation you are about to have with a doctor who has never been asked about genomics before.

The instruction manual has been inside you all along. You are just learning, perhaps for the first time, how to read it.

Figure 7: *Book One closing — typographic letterpress-style layout of the book's closing statement: 'The instruction manual has been inside you all along. You are just learning, for the first time, how to read it.' On warm cream paper with ornamental rules.*

The most important thing your genetics can give you is not a prediction. It is a reason to act — earlier, more specifically, more wisely, and with more information than any generation before you has ever had access to.

What We Covered in This Chapter

Here are the ideas and steps to carry forward:

- Open your results in the right mindset: with curiosity rather than dread, ideally with someone present, after re-reading the relevant chapters, and with your raw data already downloaded and secured.
- Phase One — immediate steps: download raw data, categories findings by type, document your family

history, and identify the one or two findings that most warrant a clinical conversation.

- Genetic counselors are the most underused resource in genomic health. They are available through NHS genetics services in the UK and through hospital departments, private practice, and telehealth in the US. Seek one out for high-penetrance findings, family planning implications, or findings that are causing significant anxiety.
- Take specific, focused questions to your GP rather than your entire report. A single focused clinical question — 'I carry a BRCA2 variant; what should we discuss about screening?' — is far more productive than an open-ended request to evaluate a hundred-page consumer report.
- Phase Two — medium term: review your medications against pharmacogenomic findings, update your screening schedule, address the family conversation, and identify your highest-priority lifestyle lever. One lever, sustained, is more valuable than ten started and abandoned.
- The lifestyle interventions with the broadest evidence base across genetic risk categories are physical activity, not smoking, sleep quality, blood pressure management, and weight management. These are not generic wellness advice — they are the specific factors that the research shows most consistently modify how genetic tendencies play out.
- Phase Three — lifetime practice: annual data re-review, annual GP conversation, living family history document, annual privacy settings review, and ongoing curiosity about the field.
- Consumer results warrant clinical follow-up for high-penetrance findings, pharmacogenomic findings affecting current medications, and VUSs in genes associated with serious conditions. A consumer test opens the conversation; a clinical test closes it with the precision medical decisions require.

BOOK ONE — YOUR DNA, YOUR BLUEPRINT — IS COMPLETE.

✦ ✦ ✦

End of Book One

Notes And Sources

Key Studies, Data Sources, And Further Reading By Chapter

This section lists the key studies and data sources that underpin specific factual claims made in each chapter and suggests further reading for readers who want to go deeper. It is not an exhaustive bibliography. Every claim in the book is either drawn from peer-reviewed research, reported by credible news organizations, or represents established scientific consensus. Where a finding is well-established and non-controversial, a source is noted but not quoted in detail. Where a finding is more recent, specific, or counterintuitive, the source is cited more fully. All URLs were verified at time of publication.

Chapter One — The Instruction Manual Inside You

Key sources

Individuals at high genetic risk for heart disease who maintained a healthy lifestyle had approximately half the cardiovascular events of those with the same genetic risk who did not.

Khera AV, Emdin CA, Drake I, et al. Genetic risk, adherence to a healthy lifestyle, and coronary disease. New England Journal of Medicine. 2016;375:2349–2358. doi:10.1056/NEJMoa1605086

You share approximately 99.9 percent of your DNA with every other human being on the planet.

International Human Genome Sequencing Consortium. Initial sequencing and analysis of the human genome. Nature. 2001;409:860–921. This figure is widely reported; the precise estimate varies slightly by methodology but is consistent across major sequencing studies.

Further reading

— Mukherjee S. The Gene: An Intimate History. Scribner, 2016. The most comprehensive lay account of genetics from Mendel to CRISPR.

— Ridley M. Genome: The Autobiography of a Species in 23 Chapters. Fourth Estate, 1999. Still the most readable introduction to what individual genes do.

— NHGRI Talking Glossary of Genomic and Genetic Terms. genome.gov/glossary — the US National Human Genome Research Institute's plain-language reference.

Chapter Two — Spitting in a Tube

Key sources

As of 2025, people of European ancestry account for more than 90 percent of all participants in genome-wide association studies.

Mills MC, Rahal C. The GWAS diversity monitor tracks diversity by disease in real time. Nature Genetics. 2020;52:242–243. The 90 percent figure is regularly updated; see also Sirugo G, Williams SM, Tishkoff SA. The missing diversity in human genetic studies. Cell. 2019;177:26–31.

The NIH All of Us Research Program has enrolled over 750,000 diverse participants and is returning personalized genetic results at no cost, with free counseling available in English and Spanish.

National Institutes of Health. All of Us Research Program. allofus.nih.gov. Program enrollment and results return milestones reported by NIH, 2024–2025.

Further reading

— Gallagher S. The DNA Detectives. Wellcome Collection, 2019. A clear account of how consumer genetic tests actually work.

— NSGC. What is genetic counseling? nsgc.org — the National Society of Genetic Counselors' plain-language overview.

Chapter Three — Reading the Risk

Key sources

Among women who carry a harmful BRCA1 variant, lifetime risk of breast cancer rises to 55–72 percent by age 70–80.

Kuchenbaecker KB, Hopper JL, Barnes DR, et al. Risks of breast, ovarian, and contralateral breast cancer for BRCA1 and BRCA2 mutation carriers. JAMA. 2017;317(23):2402–2416. doi:10.1001/jama.2017.7112

The top eight percent of cardiovascular polygenic risk scorers have likelihood of coronary artery disease more than three times higher than the general population.

Khera AV, Chaffin M, Aragam KG, et al. Genome-wide polygenic scores for common diseases identify individuals with risk equivalent to monogenic mutations. Nature Genetics. 2018;50:1219–1224. doi:10.1038/s41588-018-0183-z

People in the highest genetic cardiovascular risk tier who adopted a healthy lifestyle cut their realized ten-year risk nearly in half.

Khera AV, Emdin CA, Drake I, et al. Genetic risk, adherence to a healthy lifestyle, and coronary disease. New England Journal of Medicine. 2016;375:2349–2358. doi:10.1056/NEJMoa1605086

In populations of African descent, the dominant driver of cardiovascular disease is hypertension rather than elevated LDL cholesterol. The Heart of Soweto cohort showed Black African patients had significantly lower LDL yet comparable rates of cardiovascular disease driven primarily by blood pressure.

Sliwa K, Wilkinson D, Hansen C, et al. Spectrum of heart disease and risk factors in a black urban population in South Africa (the Heart of Soweto study): a cohort study. The Lancet. 2008;371(9616):915–922. doi:10.1016/S0140-6736(08)60417-1

About 40 percent of people who develop Alzheimer's disease do not carry the APOE e4 variant at all. Research published in 2024 in Nature Medicine suggested APOE e4 homozygotes may experience a distinct form of the disease.

Fortea J, Pegueroles J, Alcolea D, et al. APOE4 homozygosity represents a distinct genetic form of Alzheimer's disease. Nature Medicine. 2024;30:1284–1291. doi:10.1038/s41591-024-02931-w

A 2020 Lancet Commission report identified twelve modifiable risk factors accounting for around 40 percent of all dementia cases.

Livingston G, Huntley J, Sommerlad A, et al. Dementia prevention, intervention, and care: 2020 report of the Lancet Commission. The Lancet. 2020;396(10248):413–446. doi:10.1016/S0140-6736(20)30367-6

Further reading

— Topol E. The Creative Destruction of Medicine. Basic Books, 2012. Covers polygenic risk scores and precision medicine in accessible terms.

— Alzheimer's Association. alz.org — reliable plain-language information on APOE e4, risk factors, and the modifiable dementia research.

Chapter Four — Your Genes and Your Medicine Cabinet

Key sources

The PREPARE study found that genotype-guided prescribing reduced the incidence of clinically significant adverse drug reactions by 30 percent compared to standard care, across seven European countries.

Swen JJ, van der Wouden CH, Manson LE, et al. A 12-gene pharmacogenetic panel to prevent adverse drug reactions: an open-label, multicentre, controlled, cluster-randomised crossover implementation study. The Lancet. 2023;401:347–356. doi:10.1016/S0140-6736(22)01841-4

CYP2D6 ultra-rapid metabolism affects approximately 1–2 percent of people of European descent but up to 29 percent of people of North African and Ethiopian heritage.

Gaedigk A, Sangkuhl K, Whirl-Carrillo M, Klein T, Leeder JS. Prediction of CYP2D6 phenotype from genotype across world populations. Genetics in Medicine. 2017;19(1):69–76. doi:10.1038/gim.2016.80

The FDA's database of pharmacogenomic biomarkers in drug labeling currently

lists over a hundred drugs.

US Food and Drug Administration. Table of pharmacogenomic biomarkers in drug labeling. fda.gov/drugs. Updated continuously; current count verified 2025.

Further reading

— CPIC Guidelines. cpicpgx.org — the free, peer-reviewed guidelines for pharmacogenomic implementation, organized by drug and gene.

— PharmGKB. pharmgkb.org — the pharmacogenomics knowledge base, with plain-language summaries of gene–drug interactions.

— Baye TM, Abebe T, Wilke RA. Genotype–environment interactions and their translational implications. Personalized Medicine. 2011;8(1):59–70.

Chapter Five — Whole Genome Sequencing

Key sources

The Human Genome Project cost approximately three billion dollars and took thirteen years; consumer whole genome sequencing now costs under three hundred dollars.

National Human Genome Research Institute. DNA sequencing costs: data. genome.gov/about-genomics/fact-sheets/DNA-Sequencing-Costs-Data. Updated annually by the NHGRI Genome Sequencing Program.

The GRCh38 reference genome was assembled primarily from a small number of donors, with one individual of African-European origin accounting for approximately 70 percent of the sequence.

Ballouz S, Dobin A, Gillis JA. Is it time to change the reference genome? Genome Biology. 2019;20:159. doi:10.1186/s13059-019-1774-4. See also the Human Pangenome Reference Consortium. A draft human pangenome reference. Nature. 2023;617:312–324. doi:10.1038/s41586-023-05896-x

Non-European ancestry individuals typically receive approximately 23 percent more variants of uncertain significance than European ancestry individuals.

Popejoy AB, Fullerton SM. Genomics is failing on diversity. Nature. 2016;538:161–164. doi:10.1038/538161a. The 23 percent estimate is derived from multiple studies; see also Martin AR et al. Clinical use of current polygenic risk scores may exacerbate health disparities. Nature Genetics. 2019;51:584–591.

WGS delivered new diagnoses across the broadest range of rare diseases studied to date, in patients who had previously received no genetic explanation (100,000 Genomes Project).

Turro E, Astle WJ, Megy K, et al. Whole-genome sequencing of patients with rare diseases in a national health system. Nature. 2020;583:96–102. doi:10.1038/s41586-020-2434-2

In March 2025, 23andMe filed for bankruptcy; a bankruptcy judge authorized the inclusion of its customer genetic database in the asset sale process.

Multiple news reports, March–April 2025. Authoritative coverage: Molteni M. 23andMe is bankrupt. What happens to your DNA? Wired, March 2025. California Attorney General press release, March 28, 2025.

Further reading

— Davies K. The $1,000 Genome. Free Press, 2010. The story of the sequencing cost revolution, told by one of the field's leading science journalists.

— NHGRI. Genome.gov — the US National Human Genome Research Institute's public-facing resources on WGS, VUS, and the reference genome.

— ClinVar. ncbi.nlm.nih.gov/clinvar — the public database of variant classifications, updated continuously as evidence accumulates.

Chapter Six — Privacy, Data, and Who Owns Your DNA

Key sources

A genetic database covering just two percent of a population would contain enough relatives to identify the majority of its members through long-range familial searches.

Erlich Y, Shor T, Pe'er I, Carmi S. Identity inference of genomic data using long-range familial searches. Science. 2018;362(6415):690–694. doi:10.1126/science.aau4832

23andMe filed for Chapter 11 bankruptcy in March 2025. A US bankruptcy judge authorized its customer genetic database to be included in asset sale proceedings. Scholars from Harvard Law published analysis in the New England Journal of Medicine arguing existing law failed to protect genetic data.

Contreras JL, Slaughter S. Protecting genetic privacy in bankruptcy: lessons from the 23andMe case. New England Journal of Medicine. 2025. doi noted in published version. See also California Attorney General Consumer Alert, March 28, 2025.

23andMe signed a $300 million research partnership with GlaxoSmithKline in 2018.

GlaxoSmithKline and 23andMe sign agreement to leverage genetic insights for the development of novel medicines. GSK press release. July 25, 2018. gsk.com/en-gb/media/press-releases/2018.

The CODIS database significantly overrepresents people of color relative to the general population, meaning familial searching reaches families of people of color at a disproportionately higher rate.

Murphy E. Inside the Cell: The Dark Side of Forensic DNA. Nation Books, 2015. See also: Roth A. Spit and acquit: prosecutors as surveillance entrepreneurs. California Law Review. 2019;107(2):405–476.

The Golden State Killer was identified in April 2018 through forensic genetic genealogy using a consumer DNA database.

Kolata G, Murphy H. The Golden State Killer is Linked to a Crime Using DNA from a

Family Tree. The New York Times. April 26, 2018.

Further reading

— Skloot R. The Immortal Life of Henrietta Lacks. Crown, 2010. Essential reading on consent, race, and biological material in medical research.

— Reardon J. The Postgenomic Condition. University of Chicago Press, 2017. A rigorous examination of equity and diversity in genomics research.

— EFF. Electronic Frontier Foundation genetic privacy resources. eff.org/issues/genetic-information-privacy.

— Genetic Information Privacy Act tracker. Various US state laws and the Don't Sell My DNA Act: see the International Association of Privacy Professionals at iapp.org.

Chapter Seven — Your Action Plan

Key sources

The Diabetes Prevention Program demonstrated that lifestyle intervention reduced the incidence of Type 2 diabetes by 58 percent in people at high risk.

Knowler WC, Barrett-Connor E, Fowler SE, et al. Reduction in the incidence of type 2 diabetes with lifestyle intervention or metformin. New England Journal of Medicine. 2002;346(6):393–403. doi:10.1056/NEJMoa012512

150 minutes per week of moderate aerobic exercise produces clinically meaningful falls in blood pressure — particularly relevant for cardiovascular polygenic risk and for people of African ancestry where hypertension is the primary risk driver.

Cornelissen VA, Smart NA. Exercise training for blood pressure: a systematic review and meta-analysis. Journal of the American Heart Association. 2013;2:e004473. doi:10.1161/JAHA.112.004473

Cognitive engagement is associated in longitudinal studies with reduced risk of dementia and slower cognitive decline, particularly in APOE e4 carriers.

Livingston G et al. Dementia prevention, intervention, and care: 2020 report of the Lancet Commission. The Lancet. 2020;396(10248):413–446 — see modifiable risk factors section.

Further reading

— Insel TR. Healing: Our Path from Mental Illness to Mental Health. Penguin Press, 2022. On using personal health data to guide mental health decisions.

— Topol E. Deep Medicine. Basic Books, 2019. On the role of AI and genomics in the future of personalized healthcare.

— NSGC. Find a Genetic Counselor. findageneticcounselor.nsgc.org — the primary US directory for finding a certified genetic counselor.

Appendix A

Your Family Medical History

A three-generation record template

Your family medical history is the oldest genetic tool available — and one of the most powerful. Conditions that appear repeatedly across generations, or that strike at unusually young ages, are patterns your doctor needs to know about. This template gives you a structured way to record what you know and identify what to ask.

Complete what you can today. A 15-minute phone call with a parent or aunt can unlock a generation of health history. Update this document whenever a relative receives a new diagnosis. Bring it to every GP appointment alongside your genetic test results.

Free printable download

↗ **thekinnarafund.org/blueprint**

Print as many copies as needed

Condition icons

♥ Heart disease, stroke, high blood pressure, atrial fibrillation

◆ Type 2 diabetes, obesity, thyroid conditions, metabolic syndrome

▲ Cancer — note the specific type in brackets, e.g. ▲ (breast, age 54)

○ Alzheimer's, Parkinson's, dementia, neurological conditions

■ Depression, bipolar disorder, schizophrenia, anxiety disorders

✶ Autoimmune — rheumatoid arthritis, lupus, coeliac disease

? Cause unknown or not recorded

— No significant conditions recorded

Section 1 — Your own health details

About you			
Full name		**Date of birth**	
Biological sex		**Ancestry / ethnicity**	
Significant health conditions			
Current medications			
Known genetic test results			
Allergies / adverse drug reactions			

Section 2 — Parents

Mother	Father

Full name		**Full name**	
Date of birth		**Date of birth**	
Alive / deceased (age)		**Alive / deceased (age)**	
Cause of death		**Cause of death**	
Ancestry / country of origin		**Ancestry / country of origin**	
Significant conditions (icons + age)		**Significant conditions (icons + age)**	
Cancers (type + age diagnosed)		**Cancers (type + age diagnosed)**	
Notes / questions to ask		**Notes / questions to ask**	

Section 3 — Maternal grandparents (mother's parents)

Maternal grandmother		Maternal grandfather	
Full name		**Full name**	

Date of birth	**Date of birth**
Alive / deceased (age)	**Alive / deceased (age)**
Cause of death	**Cause of death**
Country of origin	**Country of origin**
Significant conditions (icons + age)	**Significant conditions (icons + age)**
Cancers (type + age)	**Cancers (type + age)**

Section 4 — Paternal grandparents (father's parents)

Paternal grandmother	Paternal grandfather
Full name	**Full name**
Date of birth	**Date of birth**
Alive / deceased (age)	**Alive / deceased (age)**

Cause of death		**Cause of death**	
Country of origin		**Country of origin**	
Significant conditions (icons + age)		**Significant conditions (icons + age)**	
Cancers (type + age)		**Cancers (type + age)**	

Section 5 — Siblings

Sibling 1		Sibling 2	
Name		**Name**	
Date of birth		**Date of birth**	
Significant conditions (icons + age)		**Significant conditions (icons + age)**	
Cancers (type + age)		**Cancers (type + age)**	

For more than two siblings, photocopy this section or continue on a separate sheet.

Section 6 — Patterns to discuss with your doctor or genetic counselor

Does any condition appear in 3+ relatives across different generations?	
Does any cancer appear in multiple relatives, or at unusually young age (under 50)?	
Do any relatives share your significant genetic test findings?	
Relatives not yet contacted / information still to gather	
Specific questions for your next GP or counselor appointment	

How to use this template

Fill in what you know today — a partial record is valuable. A short call with a parent or aunt can unlock a generation of health history.

Update this document whenever a relative receives a new diagnosis. Bring it to every GP appointment alongside your genetic test results.

Appendix B

Key Terms

Plain-English definitions of technical terms used in this book

Every technical term in this glossary is defined in context when it first appears in the main text. These definitions are here as a quick reference — a place to return when you encounter a term and want a plain-language reminder without hunting back through the chapter. Terms are listed alphabetically. The chapter where each term first appears is noted in brackets.

A

All of Us *[Ch 2]*

A research program run by the US National Institutes of Health (NIH) that has enrolled more than 750,000 participants from diverse backgrounds to build one of the most ancestry-diverse genomic databases in the world. Participants can receive personalized genetic health results at no cost. Website: allofus.nih.gov.

APOE (gene) *[Ch 2]*

A gene that produces apolipoprotein E, a protein involved in carrying cholesterol through the bloodstream. The APOE e4 variant is associated with increased lifetime risk of Alzheimer's disease. Most people who carry one copy of APOE e4 never develop Alzheimer's.

B

BAM file *[Ch 5]*

Binary Alignment Map. One of three data files produced by whole genome sequencing. A BAM file contains your raw sequencing reads reassembled and aligned against a reference genome. Typically around 60–100 GB in size. Less immediately useful for health interpretation than the VCF file, but important to store permanently.

Benign variant *[Ch 5]*

A genetic variant that has been studied in enough people to establish that it does not cause disease. The opposite of pathogenic. A benign result for a given variant is genuinely reassuring, though new evidence can occasionally reclassify variants.

BRCA1 / BRCA2 *[Ch 2]*

Two genes involved in repairing damaged DNA. Harmful variants in these genes substantially increase lifetime risk of breast, ovarian, prostate, and pancreatic cancers. BRCA variants are more common among people of Ashkenazi Jewish descent but occur across all ancestries. A BRCA flag on a consumer test should always be confirmed with clinical genetic testing.

C

Carrier *[Ch 3]*

A person who carries one copy of a recessive disease variant but does not typically develop the disease themselves. Carrier status becomes relevant for family planning — if both parents are carriers for the same recessive condition, each child has a one-in-four chance of inheriting both copies and developing the condition.

CODIS *[Ch 6]*

Combined DNA Index System. The FBI's national forensic DNA database, used for criminal identification in the United States. CODIS draws profiles predominantly from convicted offenders and arrestees and significantly overrepresents people of color relative to the general population.

Coverage (sequencing) *[Ch 5]*

How many times each position in your genome was read during whole genome sequencing. 30× coverage means each position was read independently thirty times, giving high confidence in the result. Higher coverage means greater accuracy but also larger file sizes and higher cost.

CPIC *[Ch 4]*

Clinical Pharmacogenomics Implementation Consortium. A group of international experts that publishes evidence-based guidelines for using pharmacogenomic test results to guide prescribing. CPIC guidelines are freely available and are the primary reference used by US clinicians implementing pharmacogenomic testing.

CYP enzymes *[Ch 4]*

Cytochrome P450 enzymes. A family of enzymes in the liver responsible for metabolizing approximately 70–80% of all prescribed medications. The speed at which your CYP enzymes work is controlled by your genes and varies significantly between individuals and across ancestry groups.

D

DNA (deoxyribonucleic acid) *[Ch 1]*

The molecule that carries the genetic instructions for the development, function, growth, and reproduction of all known living organisms. DNA is organised into a double helix structure made of four chemical bases — adenine (A), thymine (T), guanine (G), and cytosine (C) — whose sequence encodes all genetic information.

DPYD (gene) *[Ch 4]*

A gene that encodes an enzyme responsible for breaking down fluoropyrimidine chemotherapy drugs (5-fluorouracil, capecitabine). People with reduced-function DPYD variants cannot clear these drugs efficiently and face life-threatening toxicity at standard doses. Pre-treatment DPYD testing is now recommended by NCCN, ASCO, and CPIC in the US and is NHS policy in the UK.

E

Epigenetics *[Ch 1]*

The study of changes in gene activity that do not involve changes to the DNA sequence itself. Epigenetic changes — such as the addition of chemical tags called methyl groups — control which genes are switched on or off in different cells and at different times. Diet, exercise, stress, and sleep all influence epigenetic patterns.

Epigenome *[Ch 2]*

The complete set of epigenetic modifications across a person's genome. Unlike the genome itself, the epigenome changes throughout life in response to environment and behaviour. The epigenome is not captured by standard consumer genetic tests.

F

Familial hypercholesterolemia (FH) *[Ch 3]*

An inherited condition that causes severely elevated LDL cholesterol from birth, significantly increasing heart disease risk. Caused by variants in the LDLR, APOB, or PCSK9 genes. Affects approximately 1 in 250–300 people globally and is significantly underdiagnosed. Early identification enables clinical management that substantially normalizes cardiovascular outcomes.

FASTQ file *[Ch 5]*

The raw output file from a whole genome sequencer. A FASTQ file contains millions of short DNA sequence reads and their quality scores, before any assembly or analysis. Typically 80–90 GB in size. Not directly interpretable for health purposes but important to store as your permanent genetic archive.

G

Gene *[Ch 1]*

A specific section of DNA that carries instructions for producing a protein. The human genome contains approximately 20,000 genes, which make up roughly 2% of the total genome. The remaining 98% plays important regulatory roles that science is still working to fully understand.

Genetic counselor *[Ch 2]*

A healthcare professional with specialist training in medical genetics and counseling. Genetic counselors help people understand and interpret genetic test results, assess family history, and navigate complex decisions that findings can raise. In the US, find one at nsgc.org/findageneticcounselor.

Genome *[Ch 1]*

Your complete set of genetic instructions — all the DNA in all your cells. The human genome contains approximately three billion base pairs. Every cell in your body (with a few exceptions) contains an identical copy of your genome.

GRCh38 *[Ch 5]*

Genome Reference Consortium Human Build 38. The current standard reference genome against which your sequencing results are compared. GRCh38 was assembled primarily from a small number of donors and underrepresents the genetic diversity of non-European populations, contributing to higher rates of uncertain findings in people of non-European ancestry.

GWAS (genome-wide association study) *[Ch 2]*

A type of research study that scans the genomes of thousands of people to find genetic variants associated with particular traits or diseases. GWAS studies are the source of most polygenic risk score data. The majority of GWAS participants have historically been of European ancestry, which limits the accuracy of polygenic scores for people of other backgrounds.

H

HIPAA *[Ch 6]*

Health Insurance Portability and Accountability Act. US federal law that governs the privacy and security of health information held by healthcare providers, insurers, and their business partners. Consumer genomics companies such as 23andMe and AncestryDNA are generally not covered by HIPAA — they are governed by their own privacy policies.

High-penetrance variant *[Ch 3]*

A genetic variant that strongly and reliably predicts disease outcome. Someone who carries a high-penetrance variant has a substantially elevated lifetime risk of developing the associated condition. Examples include pathogenic BRCA1/2 variants (hereditary breast and ovarian cancer) and LDLR variants (familial hypercholesterolemia). High-penetrance variants are rare but highly actionable.

M

Metabolizer status *[Ch 4]*

A classification of how quickly your liver enzymes process a particular medication. The four categories are: poor metabolizer (very slow), intermediate metabolizer (slower than normal), normal metabolizer (standard speed), and ultra-rapid metabolizer (faster than normal). All drug dosing is calibrated for normal metabolizers.

Methylation *[Ch 7]*

The addition of a methyl chemical group (CH_3) to a section of DNA, typically at the promoter region of a gene. Methylation generally silences gene expression — it is one of the primary mechanisms through which lifestyle factors influence which genes are active at any given time.

N

NSGC *[Ch 2]*

National Society of Genetic Counselors. The professional body for genetic counselors in the United States. Their website (nsgc.org) includes a Find a Genetic Counselor tool and information on financial assistance for people without insurance coverage.

Nucleotide *[Ch 1]*

The basic building block of DNA. Each nucleotide consists of a sugar, a phosphate group, and one of four chemical bases: adenine (A), thymine (T), guanine (G), or cytosine (C). The sequence of these four bases along the DNA strand is the genetic code.

P

Pangenome *[Ch 5]*

A reference genome built from multiple individuals representing diverse human ancestries, rather than a single composite genome. The Human Pangenome Reference Consortium has developed a pangenome incorporating 47 haplotype-diverse genomes. When this becomes the clinical standard, it will substantially reduce the rate of uncertain findings for people of non-European ancestry.

Pathogenic variant *[Ch 5]*

A genetic variant that has been studied in enough people to establish that it causes or substantially increases risk of a disease. A pathogenic result is the most actionable outcome from genetic testing and should always be followed up with clinical confirmation and specialist guidance.

Penetrance *[Ch 3]*

The proportion of people who carry a particular genetic variant who actually develop the associated condition. A variant with 100% penetrance causes disease in everyone who carries it. Most disease-associated variants have incomplete penetrance — carrying the variant raises risk but does not guarantee the outcome.

Pharmacogenomics *[Ch 2]*

The study of how a person's genetic makeup affects their response to medications. Pharmacogenomic testing identifies variants in genes (primarily CYP enzymes) that affect how quickly drugs are processed by the body. Unlike disease risk prediction, pharmacogenomics applies across all ancestries, because the biochemical mechanism is universal.

Polygenic risk score (PRS) *[Ch 1]*

A number that estimates a person's genetic predisposition to a condition by adding up the effects of hundreds or thousands of small genetic variants across the genome. A high polygenic risk score does not mean a condition is inevitable — lifestyle factors can substantially modify realized risk, particularly for cardiovascular disease.

Prodrug *[Ch 4]*

A medication that is administered in an inactive form and must be converted by the body's enzymes into its active form to have an effect. Codeine and clopidogrel are prodrugs. People who are poor metabolizers of the relevant CYP enzyme may not be able to activate a prodrug effectively, meaning the medication produces little or no benefit.

Protein *[Ch 1]*

A molecule built from amino acids according to instructions encoded in a gene. Proteins are the workhorses of the body — they build structures, carry signals, catalyze chemical reactions, and defend against infection. Every function your body performs is mediated by proteins.

R

Raw data (genetic) *[Ch 2]*

The underlying data file produced by a genetic test, before it is filtered or interpreted into a consumer report. For SNP array tests, this is typically a text file containing your genotype at each tested position. For whole genome sequencing, raw data files include FASTQ, BAM, and VCF formats. You should always download and store your raw data.

Re-identification *[Ch 6]*

The process of identifying a specific individual from genetic data that was supposedly anonymous. Research has shown that genetic data from as few as 2% of a population is enough to identify the majority of that population's members through relative matching, even if names are not attached to the data.

S

SNP (single nucleotide polymorphism) *[Ch 1]*

A location in the genome where the DNA sequence varies between individuals by a single base. Pronounced 'snip.' SNPs are the most common form of genetic variation in humans. Most SNPs have no known health significance; a subset are associated with traits, disease risks, or drug responses.

SNP array *[Ch 2]*

The technology used by most consumer genetic testing companies (23andMe, AncestryDNA). A SNP array checks your DNA at up to a million pre-selected positions in the genome. It covers less than 0.1% of the genome and only checks positions pre-selected from predominantly European population studies, making it less accurate for people of non-European ancestry than whole genome sequencing.

V

VCF file (variant call format) *[Ch 5]*

The most immediately useful of the three whole genome sequencing output files. A VCF file contains only the positions in your genome that differ from the reference genome — typically around 4–5 million variants, compressed into approximately 500 MB. This is the file used for health interpretation.

Variant *[Ch 1]*

A difference between your DNA and the reference genome at a specific position. Most variants are harmless — they are simply part of normal human genetic diversity. A small number of variants are associated with increased disease risk or altered drug responses. The word 'variant' is preferred over 'mutation' in modern genetics because it carries less negative connotation.

Variant of uncertain significance (VUS) *[Ch 5]*

A genetic variant for which there is not yet enough evidence to classify it as either pathogenic (disease-causing) or benign (harmless). A VUS result is not a diagnosis and should not cause alarm. Many VUS are later reclassified as benign. People of non-European ancestry receive approximately 23% more VUS results than people of European ancestry, due to the European bias of current genomic databases.

VKORC1 (gene) *[Ch 4]*

Vitamin K epoxide reductase complex subunit 1. A gene that controls the production of clotting factors in response to warfarin. Along with CYP2C9, VKORC1 variants explain 30–40% of the variation in warfarin dose requirements between patients.

W

WES (whole exome sequencing) *[Ch 5]*

A sequencing method that reads only the protein-coding regions of the genome — the roughly 1–2% of DNA that directly encodes genes. WES is less comprehensive than WGS but more targeted and less expensive. It is used clinically for rare disease diagnosis but has largely been superseded by WGS as sequencing costs have fallen.

WGS (whole genome sequencing) *[Ch 5]*

A method that reads all three billion base pairs of your DNA rather than just pre-selected positions. Standard consumer WGS uses 30× coverage (each position read thirty times) and produces three output files: FASTQ, BAM, and VCF. WGS is the most comprehensive and ancestry-equitable form of genomic testing currently available to consumers.

Terms are defined as used in this book. Scientific understanding evolves — for the latest definitions consult a current clinical genetics resource or your genetic counselor.

Appendix C

Finding A Genetic Counselor

Resources for English-speaking countries

Genetic counselors are the most important professional resource this book can point you toward. They are specialists in translating genomic findings into plain language, helping you understand what your results mean in the context of your personal and family history, and guiding you toward appropriate clinical follow-up. If you have received a significant genetic finding or are simply uncertain what to do with your test results, a genetic counselor is the right first call.

The resources below are current as of publication. Website addresses and phone numbers occasionally change — if a link is not working, search the organization name directly. All organizations listed are professional bodies or established clinical services.

> **Before you search — three things to know**
>
> 1. Telehealth has transformed access. You no longer need to be near a major hospital to see a genetic counselor. Most of the organizations below offer video or phone consultations that work from anywhere.
>
> 2. Insurance coverage varies. In the US, many plans cover genetic counseling when ordered by a physician. Ask your GP to make the referral before booking directly — this often brings the cost within insurance scope.
>
> 3. If cost is a barrier, say so. Many genetic counselors offer sliding-scale fees for uninsured patients. Community health centers increasingly have genetics referral pathways. The organizations below can advise on financial assistance where it exists.

United States

National Society of Genetic Counselors (NSGC)

Find a counselor: findageneticcounselor.nsgc.org

Main website: nsgc.org

Phone: (312) 321-6834

Searchable directory of over 3,300 certified genetic counselors in the US and Canada. Search by location, specialty, or telehealth availability. The NSGC is the primary professional body for genetic counselors in the US.

American Board of Genetic Counseling (ABGC)

Website: abgc.net

The credentialing body that certifies genetic counselors in the US. Use the ABGC directory to verify that a counselor holds current board certification (CGC credential).

Genome Medical — Telehealth (US and Canada)

Website: genomemedical.com

Appointments: Available within one week, all 50 states plus Canada

Self-pay cost: $250 per session (insurance often covers)

The largest telehealth genetic counseling provider in the US. Board-certified genetic counselors available by phone or video with no GP referral required. Covers hereditary cancer, pharmacogenomics, rare disease, cardiology, and more. Delivered its 250,000th session in June 2025.

Color Genomics — Equity-Focused Clinical Genetics

Website: color.com

Offers clinical genetic testing and counseling through employer health plans and community health programs, often at low or no cost. Official counseling partner for the NIH All of Us program. Specifically designed to reach communities with limited access to traditional genetics services.

Federally Qualified Health Centers (FQHCs)

Find a center: findahealthcenter.hrsa.gov

Community health centers that provide primary care on a sliding-scale basis regardless of insurance status. Increasingly have genetics referral pathways. The most accessible entry point into the US healthcare system for those without insurance.

NIH All of Us Research Program

Website: allofus.nih.gov

Genetic counseling: Free for enrolled participants, in English and Spanish

Research program enrolling diverse participants to build a more representative genomic database. Participants receive personalized genetic health results and free genetic counseling. Open to all US residents aged 18 and over.

Canada

Canadian Association of Genetic Counsellors (CAGC)

Find a counselor: cagc-accg.ca

The national professional body for genetic counselors in Canada. Searchable directory by province and specialty. Also provides certification of Canadian genetic counselors through examination.

Canadian Board of Genetic Counselling (CBGC)

Website: cbgc-cccg.ca

The credentialing body for genetic counselors in Canada. Use the CBGC to verify that a counselor holds the CCGC (Certified Canadian Genetic Counsellor) credential.

Genome Medical — Telehealth

Website: genomemedical.com

Coverage: All Canadian provinces via telehealth

See United States entry above for full details.

United Kingdom and Ireland

NHS Genomic Medicine Service

Website: england.nhs.uk/genomics

Access: Via GP referral to NHS clinical genetics services

Clinical genetics services are available across all NHS regions. Ask your GP to refer you to your regional NHS Genomic Medicine Centre. The NHS does not charge for genetics consultations. Waiting times vary by region and clinical urgency.

Association of Genetic Nurses and Counsellors (AGNC)

Website: agnc.org.uk

The professional organization representing genetic counsellors, genetic nurses, and patient-facing genetics staff in the UK and Ireland. Provides a public directory of registered practitioners and information on standards of practice.

Genetic Counsellor Registration Board (GCRAB)

Website: registration.agnc.org.uk

The voluntary registration body for genetic counsellors in the UK and Republic of Ireland. Use the register to verify that a counsellor has met the professional competency standards. The profession is moving toward statutory regulation through the Health and Care Professions Council (HCPC).

Genetic Alliance UK

Website: geneticalliance.org.uk

Phone: 020 7704 3141

A patient organization representing people affected by genetic conditions. Provides guidance on accessing NHS genetics services and supports families navigating the genetics system.

Australia and New Zealand

Human Genetics Society of Australasia (HGSA)

Website: hgsa.org.au

Find a counselor: hgsa.org.au/Web/Consumer-resources/Genetic-Counselling.aspx

The peak professional and scientific body for human genetics in Australia and New Zealand. The HGSA certifies genetic counselors and provides a public directory. Certified counselors hold the MHGSA credential; senior practitioners hold FHGSA.

Australasian Society of Genetic Counsellors (ASGC)

Website: asgc.org.au

The professional society representing over 500 genetic counselors across Australia and New Zealand. A special interest group of the HGSA. Provides resources for people seeking genetic counseling and maintains a network of practitioners.

Genetic Support Network of Victoria

Website: gsnv.org.au

Phone: 1800 443 078 (Australia-wide)

Provides information and support for individuals and families affected by genetic conditions across Australia, including guidance on accessing genetics services in each state and territory.

South Africa

Genetic Counselors of South Africa (GCSA)

Website: gcsa.org.za

The professional body for genetic counselors in South Africa, operating as a focus group of the South African Society of Human Genetics. Works with the Health Professions Council of South Africa (HPCSA), which is the statutory regulatory body for genetic counselors in South Africa.

Health Professions Council of South Africa (HPCSA)

Website: hpcsa.co.za

The statutory regulatory body for all healthcare professionals in South Africa including genetic counselors. Use the HPCSA register to verify a counselor's registration status.

Division of Human Genetics, University of the Witwatersrand

Website: wits.ac.za

Location: Johannesburg, South Africa

One of the leading genetics clinical and research centers in Africa. Provides clinical genetics services and is a referral center for complex cases. Ask your GP for a referral.

Telehealth — accessible from any English-speaking country

Genome Medical

Website: genomemedical.com

Languages: English

Coverage: All US states, Canada; international consultations available

No GP referral required. Appointments within one week. Self-pay $250; insurance often covered for US patients.

Invitae Genetic Counseling

Website: invitae.com

Clinical-grade genetic testing company that offers genetic counseling sessions alongside its testing services. Accessible from multiple countries. Particularly strong in hereditary cancer, cardiovascular, and rare disease.

A note on this list

Websites, phone numbers, and service details change. If any link in this appendix is not working, search the organization name directly. All organizations listed were verified as current and active at the time of publication. This list covers the countries where most English-speaking readers are likely to be based. If your country is not listed, the NSGC's international directory (nsgc.org) and the Transnational Alliance for Genetic Counseling (tagcgc.org) are the most comprehensive global resources.

About The Author

Debra-Ann Smith

M.S. Biotechnology Enterprise & Entrepreneurship, Johns Hopkins University · B.A. Harvard University, Minor in Biotechnology

Debra-Ann Smith has spent her career doing one thing with exceptional skill: finding the clearest possible way to communicate something complicated to people who need to understand it. As an award-winning documentary producer — her work includes the Telly Award-winning film Tangles & Locks and seven documentaries for the Smithsonian-Showtime Channel — she learned early that no communication tool is off the table when science needs to reach a general audience.

After earning a Bachelor of Arts from Harvard University, that conviction deepened when she returned to academia as an adult. She completed a Master of Science in Biotechnology Enterprise and Entrepreneurship at Johns Hopkins University in 2018, and a Certificate in the Ethics of AI from the London School of Economics in 2023. What she found in graduate school confirmed what she had always suspected: the most important scientific breakthroughs of our time were reaching the public too late, too incompletely, and in language that excluded more people than it included.

In response, she founded the Kinnara Fund, a Washington D.C.-based nonprofit whose mission is to make science accessible and to advance the principles of bioethics in public, institutional, and corporate life. The Fund's name comes from mythology: a Kinnara is a being that is part human, part bird — its role is to guard the well-being of humans in times of trouble. Raising awareness, Debra-Ann believes, is exactly that.

The Biology of You grew directly from the Fund's mission. The public whose tax dollars fund academic research into genomics, pharmacogenomics, and the ethics of biological data deserves to understand what that research means for their own health — not in ten years, but now. Debra-Ann believes that a well-informed public is not just better for individuals: it is better for society. Informed citizens ask better questions of their doctors, make more grounded decisions about genetic testing, and participate more meaningfully in the policy conversations that shape how biotechnology develops.

Alongside her work at the Kinnara Fund, Debra-Ann is co-owner of The Arima Group, a New York-based business management consultancy with more than twenty years of experience in the technology and lifestyle industries. She has produced acclaimed Off-Broadway theatre, executive produced documentary films, and co-founded Pretty Pretty Cosmetics, which launched at Fred Segal in Los Angeles and was featured in Vogue, Harper's Bazaar, and Elle.

The Biology of You, Your DNA, Your Blueprint is her first book. It will not be her last.

About The Kinnara Fund

> *In mythology, a Kinnara is part human, part bird — head, torso, and arms of a person, with the wings, tail, and feet of a bird. Their role is to guard the well-being of humans in times of trouble or danger. They raise awareness. That is what this Fund exists to do.*

The Kinnara Fund is a Washington D.C.-based nonprofit founded by Debra-Ann Smith and Homer K. Sun. Its mission is to make science accessible — to demystify the biotechnology breakthroughs that are reshaping medicine, privacy, and human health, and to put that knowledge directly into the hands of the citizens whose tax dollars fund the research.

The Fund was born from a straightforward observation: the gap between what academic science produces and what the public understands is not a gap in intelligence. It is a gap in communication. The journals, the conferences, and the laboratory announcements are not written for the person who will eventually be affected by what they describe. That person is everyone.

Debra-Ann Smith brings more than two decades of enterprise technology leadership, a career as an award-winning documentary producer, and graduate training in biotechnology enterprise and entrepreneurship at Johns Hopkins University. Homer K. Sun brings four decades of systems engineering and technology architecture experience spanning defense systems, Fortune 500 infrastructure, and digital transformation across industries. Between them, the founders have spent their careers doing what the Kinnara Fund does: building things that work in the real world, for real people, grounded in evidence and guided by ethics.

The Fund's work spans three areas. The first is scientific communication — producing books, educational platforms, podcasts, and public programs that translate complex science into language that serves a general audience. The Biology of You is the first expression of this work in print. The second is bioethics advocacy — engaging legislators, institutions, and corporate leaders on the ethical implications of biotechnology, genetic data, and artificial intelligence, and working to ensure that the governance frameworks shaping these technologies are informed by public values, not just industry priorities. The third is responsible technology practice — helping organizations navigate AI and technology adoption through a framework the Fund calls the Skeptic-First Approach: problem before tool, ethics before deployment, evidence before commitment.

The Fund's position is not anti-technology. It is anti-haste and anti-harm. Every field covered in this book — genomics, pharmacogenomics, genetic privacy, whole genome sequencing — holds extraordinary promise. That promise is best realized when the public understands what is happening, can ask informed questions of their doctors and their elected representatives, and can give genuinely informed consent to the decisions that affect their lives.

A well-informed public is not just better for individuals. It is better for science, better for policy, and better for society.

The Kinnara Fund · Washington D.C.

thekinnarafund.org · biotethic.com

The Biology Of You

An eight-book series on biotechnology and your health

Each book in this series takes one area of biotechnology and explores it entirely from your perspective — what it means for your health, what is available to access today, and how to use it wisely.

Book 1 Your DNA, Your Blueprint · This volume

Understanding your genome, your health risks, and the power you hold to shape your own biology

Book 2 The Universe Inside You · Coming soon

Your microbiome — the 38 trillion microbes shaping your immunity, mood, and metabolism

Book 3 The Spark Inside You · Coming soon

Bioelectricity — the electrical signals your body uses to heal, grow, and defend itself, and what they mean for the future of medicine

Further books in the series will cover:

— *Precision medicine and the therapeutics revolution*

— *The science of longevity and biological aging*

— *CRISPR, gene editing, and what the age of biological rewriting*

— *Personalized nutrition and metabolic health*

— *Regenerative medicine and the future of recovery*

— *Artificial intelligence, wearables, and your health data*

Visit thekinnarafund.org/books for updates on new releases.

A Small Request

If this book was useful to you, a short review on Amazon or Goodreads makes an enormous difference. It takes two minutes, and for an independent author it matters more than you might expect — reviews are how other curious readers find their way to books like this one.

The Biology of You series was written to put serious science into the hands of people who don't have a science degree but do have a stake in their own health. That includes a lot of people who will never encounter this book through conventional channels.

If you found it valuable, consider sharing it — with a family member who has had a genetic test and doesn't quite know what to do with the results, with a friend who has been putting off looking into their health history, or with a community that deserves access to this kind of information as much as anyone else.

That is the whole point of the Kinnara Fund: that the biology of you should belong to you, regardless of where you started.

* * *

The Kinnara Fund · Washington D.C.

ACKNOWLEDGEMENT

No man is an Island.

Homer, Nahshon, Leila, Nadine, MaryAnn, Judy, Sarah, Xioping, Emmanuel, Keston, Jamil, Sheena, Asher, Aaron, Brent, Zulema, Vanette, Wendy, Grace, Trevor, Paul, Susan, Patricia, Shirley.

Thank you for seeing me through this chapter.

www.ingramcontent.com/pod-product-compliance
Lightning Source LLC
LaVergne TN
LVHW052316200926
848481LV00002B/585